AF352359

Fred Gage   Yves Christen (Eds.)

Retrotransposition, Diversity and the Brain

With 31 Figures, 21 in color

 Springer

Gage, Fred H., Ph.D.
Laboratory of Genetics
The Salk Institute for Biological Studies
10010 North Torrey Pines Road
La Jolla, CA 92037
USA
e-mail: gage@salk.edu

Christen, Yves, Ph.D.
Fondation IPSEN
Pour la Recherche Thérapeutique
24, rue Erlanger
75781 Paris Cedex 16
France
e-mail: yves.christen@ipsen.com

ISSN 1861-2253
ISBN 978-3-540-74965-3 Springer Berlin Heidelberg New York

Cataloging-in-Publication Data applied for Bibliographic information published by Die Deutsche Bibliothek
Die Deutsche Bibliothek lists this publication in the Deutsche Nationalbibliografie; detailed bibliographic data is available in the Internet at <http://dnb.ddb.de>.

Springer is a part of Springer Science+Business Media
springer.com

Cover design: WMXDesign GbmH, Heidelberg, Germany
Typesetting and production: LE-TEX Jelonek, Schmidt & Vöckler GbR, Leipzig, Germany
Printed on acid-free paper   27/3100/YL   5 4 3 2 1 0
SPIN 12123797

Preface

The human brain is remarkably complex, permitting sophisticated behavioural repertoires, such as languages, tool use, self-awareness, symbolic thought, cultural learning and consciousness. Each human being is different, due in part to the uniqueness of the neuronal heterogeneity and interconnections in our brains. Brain complexity and neuronal diversity are strongly related. The diversity of single neurons provides the underpinnings for how neuronal circuits operate. How and when neuronal diversity is generated, both in embryonic and adult neurogenesis, remain unknown.

In the immune system, the highly diverse array of antigen receptors can be attributed to the stochastic nature of the recombination process in somatic precursor cells, causing permanent changes in DNA and gene expression. This diverse population is then the target of selective processes that favor the correct antigen-receptor match and eliminate those with inadequate specificities, accounting for the rapid kinetics and immense diversity observed *in vivo*. Evidence for a possible similarity between the nervous and immune systems came from studies with mice deficient in DNA double strand break (DSB) repair. Lessons learned from the discovery of the mechanism for diversity in the immune system may be useful to the investigation of the mechanism of diversity in neurons.

Retroelements are ancient mobile DNA found in most organisms. Long dismissed as useless, selfish or "junk" DNA, they were thought to be mere intracellular parasites from our distant evolutionary past. Together with their mutant relatives, L1 sequences constitute almost 50% of the mammalian genome. L1s can retrotranspose in a defined window of the neuronal differentiation, changing the genetic information in single neurons in a "random" fashion, allowing the brain to develop in distinct different ways. Such strategy contributes to expand the number of functionally distinct neurons that could be produced from a given stem cell gene pool. This characteristic of variety and flexibility may contribute to the uniqueness of an individual brain, even between genetically identical twins. These mobile elements may be part of conserved core process responsible for evoking facilitated complex non-random phenotypical variation on which selection may act. A detailed understanding of the basic mechanisms of L1 activity may shed light on one possible mechanism for generating neural diversity.

This Fondation IPSEN Colloque Médecine et Recherche was devoted to the interface between the complexity of brain organization and function, the mechanisms for generating diversity and genetic mobility. The goal was to expand the current limits of research in neurobiology not only to the benefit of those interested in the cellular and molecular processes but also for the understanding of high-level cognitive functions and the understanding of complex mental diseases.

Fred Gage
Yves Christen

Table of Contents

Telomeres and Telomerase in Human Health and Disease
J. Lin, E.S. Epel, E.H. Blackburn 1

Molecular and Circuit Mechanisms for Hippocampal Learning
S. Tonegawa, T.J. McHugh 13

Retrotransposons – Natural and Synthetic
J.D. Boeke, W. An, L. Dai, E.S. Davis, J.S. Han, K.A. O'Donnell, L.Z. Scheifele,
S.J. Wheelan 21

Ancient Retrotransposons as Possible Remnants of the Primitive RNP World
R. Ivanyi-Nagy, J.-L. Darlix 33

Human Diversity and L1 Retrotransposon Biology: Creation of New Genes
and Individual Variation in Retrotransposition Potential
H.H. Kazazian, Jr., M.d.C. Seleme, D.V. Babushok, D.M. Ostertag, M.R. Vetter,
P.K. Mandal 43

From the "RNA World" to Brain Complexity: Generation of Diversity
A.R. Muotri, M.C.N. Marchetto, F.H. Gage 53

Endogenous Retroviruses and Human Neuropsychiatric Disorders
R.H. Yolken, H. Karlsson, I. Bossis, L. Asp, F. Dickerson, C. Nellåker, M. Elashoff,
E. Rubalcaba, R.P. Viscidi 65

Is Psychosis Due to Retroviral/Retrotransposon Integration Close
to the Cerebral Dominance Gene?
T.J. Crow, J.S. Close, H.-S. Kim, M.T. Ross 87

Microcephalies and DNA Repair
E.C. Gilmore, C.A. Walsh 109

Subject Index 121

List of Contributors

Asp, Linnéa
Department of Neuroscience, Karolinska Institutet, Retzius v 8, 17177 Stockholm, Sweden

Babushok, D.V.
Department of Genetics, University of Pennsylvania School of Medicine, Philadelphia, PA 19104, USA

Blackburn, Elizabeth H.
Biochemistry and Biophysics, University of California, San Francisco, CA 94158, USA

Boeke, Jef D.
High Throughput Biology Center, Johns Hopkins University School of Medicine, Baltimore, MD 21205, USA

Bossis, Ioannis
Stanley Laboratory of Developmental Neurovirology, Johns Hopkins University School of Medicine, Baltimore, MD, USA

Close, J.S.
Sane Powic, Warneford Hospital, Oxford, OX3 7JX, USA

Crow, Timothy J.
Sane Powic, Warneford Hospital, Oxford, OX3 7JX, UK

Dai, Lixin
High Throughput Biology Center, Johns Hopkins University School of Medicine, Baltimore, MD 21205, USA

Darlix, Jean-Luc
LaboRetro, Unité de Virologie humaine INSERM, IFR128, ENS Lyon 46 allée d'Italie, 69364 Lyon, France

Davis, Edward S.
High Throughput Biology Center, Johns Hopkins University School of Medicine, Baltimore, MD 21205, USA

Dickerson, Faith
Sheppard Pratt Health System, Baltimore, MD, USA

Elashoff, Michael
Stanley Medical Research Institute, Chevy Chase, MD, USA

Epel, Elissa S.
Department of Psychiatry, UCSF Health Psychology Program, San Francisco, CA 94143, USA

Gage, Fred H.
Laboratory of Genetics, The Salk Institute for Biological Studies, 10010 North Torrey Pines Road, La Jolla, CA 92037, USA

Gilmore, Edward C.
Department of Neurology, Beth Israel Deaconess Medical Center, Child Neurology, Massachusetts General Hospital, MA, USA

Han, Jeffrey S.
High Throughput Biology Center, Johns Hopkins University School of Medicine, Baltimore, MD 21205, USA

Ivanyi-Nagy, Roland
LaboRetro, Unité de Virologie humaine INSERM, IFR128, ENS Lyon 46 allée d'Italie, 69364 Lyon, France

Karlsson, Håkan
Department of Neuroscience, Karolinska Institutet, Retzius v 8, 17177 Stockholm, Sweden

Kazazian, H.H., Jr.
Department of Genetics, University of Pennsylvania School of Medicine, Philadelphia, PA 19104, USA

Kim, Heui-Soo
Section of Biological Systems, College of Natural Science, Pusan National, University, San 30, Changjeon Dong, Pusan 609-735, South Korea

Lin, Jue
Biochemistry and Biophysics, University of California, San Francisco, CA 94158-2517, USA

Mandal, P.K.
Department of Genetics, University of Pennsylvania School of Medicine, Philadelphia, PA 19104, USA

Marchetto, Maria C.N.
Laboratory of Genetics, The Salk Institute for Biological Studies, 10010 North Torrey Pines Road, La Jolla, CA 92037, USA

McHugh, Thomas J.
The Picower Institute for Learning & Memory, RIKEN-MIT Neuroscience Research Center, Massachusetts Institute of Technology, 77 Massachusetts Avenue, Cambridge MA 02139-4307, USA

Muotri, Alysson R.
Laboratory of Genetics, The Salk Institute for Biological Studies, 10010 North Torrey Pines Road, La Jolla, CA 92037, USA

Nellåker, Christoffer
Department of Neuroscience, Karolinska Institutet, Retzius v 8, 17177 Stockholm, Sweden

O'Donnell, Kathryn A.
High Throughput Biology Center, Johns Hopkins University School of Medicine, Baltimore, MD 21205, USA

Ostertag, D.M.
Department of Genetics, University of Pennsylvania School of Medicine, Philadelphia, PA 19104, USA

Ross, M.T.
X Chromosome Group, Wellcome Trust Sanger Institute, Wellcome Trust Genome Campus, Hinxton, Cambridge CB10 1SA, USA

Rubalcaba, Elizabeth
Stanley Laboratory of Developmental Neurovirology, Johns Hopkins University School of Medicine, Baltimore, MD, USA

Scheifele, Lisa Z.
High Throughput Biology Center, Johns Hopkins University School of Medicine, Baltimore, MD 21205, USA

Seleme, M.d.C.
Department of Genetics, University of Pennsylvania School of Medicine, Philadelphia, PA 19104, USA

Tonegawa, Susumu
The Picower Institute for Learning & Memory, RIKEN-MIT Neuroscience Research Center, Massachusetts Institute of Technology, 77 Massachusetts Avenue, Cambridge MA 02139-4307, USA

Vetter, M.R.
Department of Genetics, University of Pennsylvania School of Medicine, Philadelphia, PA 19104, USA

Viscidi, Raphael P.
Stanley Laboratory of Developmental Neurovirology, Johns Hopkins University School of Medicine, Baltimore, MD, USA

Walsh, Christopher A.
Department of Neurology, Beth Israel Deaconess Medical Center, Division of Genetics, Children's Hospital Boston, Howard Hughes Medical Institute, Boston, MA, USA

Wenfeng, An
High Throughput Biology Center, Johns Hopkins University School of Medicine, Baltimore, MD 21205, USA

Wheelan, Sarah J.
High Throughput Biology Center, Johns Hopkins University School of Medicine, Baltimore, MD 21205, USA

Yolken, Robert H.
Stanley Laboratory of Developmental Neurovirology, Johns Hopkins University School of Medicine, Baltimore, MD, USA

Telomeres and Telomerase in Human Health and Disease

Jue Lin[1], *Elissa S. Epel*[2], and *Elizabeth H. Blackburn*[1]

Introduction

Telomeres cap chromosome ends and help protect the genome. Telomere maintenance consists of an integrated cellular system for telomere homeostasis that includes telomerase, which replenishes telomeric DNA lost from chromosomal termini. Telomerase, with its highly specialized reverse transcriptase action, is therefore essential for genomic stability and long-term cell division. The activity of telomerase in human cells is kept under a complex set of controls that include developmental, cell type-specific and environmental modulators. We have reported that chronic psychological stress in people leads to lower telomerase and shorter telomeres. From these and other studies, the emerging overall pattern is that telomerase insufficiency is associated with conditions, syndromes and diseases that can shorten human life.

Telomeres

Telomeres are DNA-protein complexes at the ends of eukaryotic chromosomes that are essential for genomic stability. The telomeric complexes prevent the ends of linear chromosomes from being recognized as broken ends, which would otherwise elicit inappropriate DNA damage responses with potentially deleterious consequences (Blackburn 2001). Telomeric DNA sequences are lost after each cell division due to incomplete replication by conventional DNA polymerases. Such progressive loss of telomeric sequences – due to incomplete replication of DNA, and potentially also from nuclease action on telomeric termini – leads to replicative senescence of dividing cells.

Telomerase: a Specialized Cellular Reverse Transcriptase Essential for Continued Cell Renewal

This end-replication problem is solved by the cellular enzyme telomerase. Telomerase, a specialized ribonucleoprotein reverse transcriptase, synthesizes telomeric DNA, thus

[1] Biochemistry and Biophysics, University of California, San Francisco, CA 94158-2517, USA
   e-mail: elizabeth.blackburn@ucsf.edu
[2] Department of Psychiatry, UCSF Health Psychology Program, San Francisco, CA 94143, USA

Gage et al.
Retrotransposition, Diversity and the Brain
© Springer-Verlag Berlin Heidelberg 2008

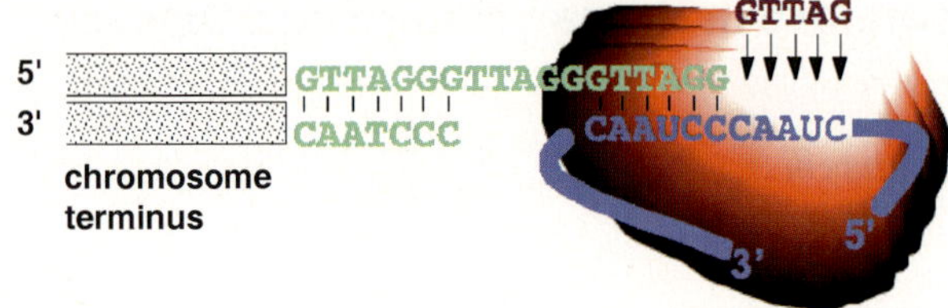

Fig. 1. A simplified diagram depicting human telomerase acting to elongate a chromosomal DNA end. Deoxyribonucleoside triphosphate substrates (with base identities indicated in *purple*) are added, templated by the RNA bases shown in *blue*. The core protein component hTERT is shown in *red* and the essential telomerase RNA component hTER in *blue*

counteracting the losses of telomere sequence (Greider and Blackburn 1985). For this purpose, telomerase uses its integral RNA molecule as the template to synthesize telomeric sequence DNA (Greider and Blackburn 1987; 1989). The core telomerase complex contains two subunits that are essential for its catalytic function: the protein catalytic subunit (hTERT in humans; Nakamura et al. 1997) and the RNA component (hTER, hTR or hTERC in humans; Greider and Blackburn 1989; Feng et al. 1995; Counter et al. 1997; Lingner et al. 1997; Fig. 1). Like the reverse transcriptase (RT) of the human retroelement LINE-1 (Piskareva and Schmatchenko 2006), telomerase lacks an RNase H activity. Also like the human LINE element reverse transcriptase (Kulpa and Moran 2006), telomerase copies the RNA that is within the same telomerase RNP complex as the protein RT subunit, not an RNA template added in trans. As discussed below, in humans, telomerase activity is regulated during development and by different environmental and physiological factors (Aisner et al. 2002; Cong et al. 2002; Forsyth et al. 2002).

The Evolution of Telomerase

Telomerase is found throughout eukaryotes ranging from those in deep branches in the eukaryotic phylogeny (for example, *Giardia* and *Trypanosoma*) to protozoans, fungi, plants, and the invertebrate and vertebrate metazoans. Hence telomerase is likely to have been present early in the eukaryotic lineage or even at the very onset of that lineage. But where did telomerase come from? How is it related to other reverse transcriptases, such as those of retroelements (Fig. 2A)? Telomerase is conserved in multiple ways: in being an RNP enzyme, in having domains of TERT conserved in addition to its RT domain, and in having a conserved core structure of the telomerase RNA moiety (Fig. 2B). Two different types of models have been considered for the evolution of telomerase. These models, described next, are not mutually exclusive.

The "Catalytic RNA to Telomerase RNP" Model

The first model, which will be called the "RNA to telomerase RNP" model, was put forth based on studies of the telomerase RNA component, the first of the core components to

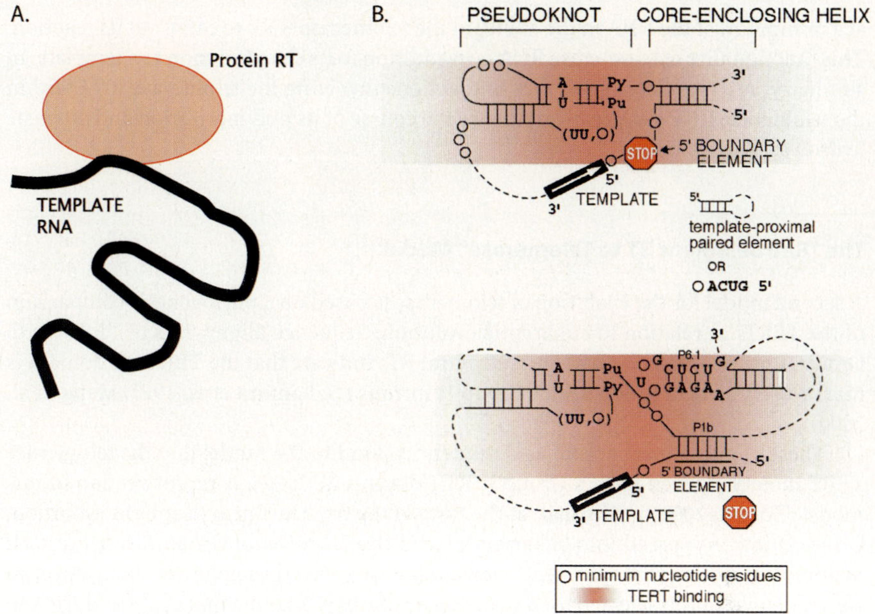

Fig. 2. A. A typical retroelement reverse transcriptase (*oval*) and its long template RNA (*black line*). **B.** The conserved core structure of telomerase RNA; note that only a limited region of the RNA is used as a template for DNA synthesis. *Top*: TER of yeasts and ciliates; *bottom*, TER of vertebrates (Lin et al. 2004)

be identified and functionally and structurally analyzed (Blackburn 1993, 1999). This model was inspired by the discovery that RNA can act as a biological catalyst, as it does in the case of self-splicing introns, RNAse P and the ribosome, and possibly in general pre-mRNA splicing. Catalytic RNAs have been constructed with templated nucleic acid polymerization properties. Hence, one model is that telomerase might have evolved from an ancestral catalytic RNA that acquired, at some point in its evolution, a protein component that took over RT catalytic function from the RNA.

The RNA to telomerase RNP model was proposed because, although it was clear that protein is an essential part of telomerase and contains catalytic site amino acid residues, certain small mutations of telomerase RNA residues often caused drastic effects on telomerase function in vivo and in vitro. Striking examples have been observed in multiple eukaryotes' telomerases over several years of research. They include small base substitutions that led to quantitatively large and RNA mutant-specific effects on the rates of dNTP misincorporation, template slippage and mis-alignment on the template (reviewed in Blackburn 1999; Lin et al. 2004). In some cases, even single base substitutions led to deleterious and large effects on the enzyme reaction. In addition, the telomeric DNA bases, even those that do not base-pair with the template, interact with the TERT moiety of telomerase, and such interaction can have a large effect on the catalytic reaction rate itself. This effect occurs at a step other than the DNA or dNTP substrate binding steps, or the product release step of the polymerization reaction (Lee and Blackburn 1993; Lee et al. 1993). Together these results point to a close involvement, at a minimum, of the RNA in the ability of the telomerase RNP to carry out its reaction. This functionality of telomerase RNA is in addition to its clear function as a template. In summary, such results suggest that interactions involving the telomerase RNA within the telomerase RNP can greatly influence the course of its polymerization and intrinsic hydrolysis reactions.

The "Retroelement RT to Telomerase" Model

A second model for the evolution of telomerase is based on a phylogenetic comparison of the TERTs in relation to eukaryotic evolution. Sequence alignments of TERTs with Group II intron, retroelement and retroviral RTs indicate that the TERT RT domain is most closely related to the RTs of Group II introns (Nakamura et al. 1997; Malik et al. 2000)

These alignments of amino acid sequences point to the model that the telomerase TERT moiety evolved from a Group II RT (Malik et al. 1997). A recent version of this model (Koonin 2006) posits that, at the time of the Archea-Eukarya split in evolution, Group II introns spread into eukaryotes (with this invasion of Group II introns itself possibly promoting that split), and that telomerase evolved when the ensuing chromosome fragmentation selected for a Group II RT-derived enzyme that could add DNA to chromosome ends. Thus, in this model, an ancestral Group II RT protein evolved into telomerase.

To complete any account of the inferred origin of telomerase, we must consider how it acquired, for its now specialized DNA addition role, the built-in template that is included within the larger telomerase RNA molecule. As described above, in the "catalytic RNA to telomerase RNP" model, an ancestral, originally catalytically competent telomerase RNA acquired the RT that then diverged into Group II RTs and into the TERT moiety of telomerase. In the alternative "retroelement RT to telomerase" model, this ancestral Group II intron RT, which had a conserved active site containing metal-coordinating aspartates mediating catalysis, acquired an RNA. One conceivable partial reconciliation of the two classes of models is that the telomerase RNA that was acquired by a common RT protein ancestor of Group II introns and TERT was derived from a ribozyme that had properties particularly suited to the role of adding tandem short repeats specifically to telomeres. Such properties could include its tight and specific binding to the TERT so that there was no dissociation of the template or its DNA product from the RT after each round of copying the short template. Consistent with this idea, much of the conserved core structure of telomerase RNA is involved in its specific binding to the TERT protein in a way that promotes optimal and specific template usage. This model leaves open the question of whether a direct ancestor of telomerase RNA ever had any nucleic acid polymerization catalytic capability.

Control of telomerase in human cells

Although the controls of telomerase activity are many-faceted and complex, some generalizations may be made. Telomerase activity is high in mammalian embryonic stages but is decreased later in life (Wright et al. 1996). Indeed, the majority – although not all – of human somatic cell types have undetectable or very low telomerase activity. However, the importance of telomerase, albeit at low levels, is becoming increasingly evident in multiple human cell types, such as resting white blood cells and fibroblasts. As a result of critically short telomeres resulting from the long-term insufficiency of telomerase for telomere maintenance, such cells can enter replication senescence (Harley et al. 1990). Telomere shortening was thus thought of as an unopposed mitotic clock that counts the number of divisions a cell is able to go through before senescence (Harley et al. 1990). However, although low, telomerase activity is expressed in a highly regulated manner in some somatic cells. For example, during lymphocyte development, differentiation and activation, telomerase activity is high in early stages of T and B cell development, but the activity is decreased at later stages and in resting cells, although it can be measured with suitable quantitative methods (Epel et al. 2004, 2006). These findings imply that, in any cells with even low telomerase, the rate of telomere shortening can be modulated by, among other factors, the telomerase activity that counteracts such shortening. With respect to the brain, formerly thought to have no telomerase and to be essentially comprised of postmitotic cells, we have found low telomerase activity in rodent hippocampus that includes stem and neural progenitor cells and have also detected low telomerase activity in primary (that is, non-transformed) human neurons in culture (J.L. and E.H.B, unpublished observations). Dividing brain stem cells have been recently reported in human adults and presumably these cells will also contain telomerase activity.

Multiple studies on aspects of telomerase control in cultured human cells have been done (see, as representative examples, Endoh et al. 2005; Ritz et al. 2005). The levels of telomerase core components TERT and TER, and of enzymatic activity, are controlled by transcriptional control. Cis-acting elements in the TERT and TER promoters include both positive and negative controlling elements. In addition, various post-transcriptional control mechanisms resulting in regulation of telomerase activity have been described for certain mammalian cells. However, much research remains to be done to understand fully the control of telomerase expression and its activity in any tissue, let alone in the mammalian or human brain.

Other Roles of Telomerase Besides Making Telomeric DNA Longer

Evidence is building for a cellular response to telomerase status independent of its role in polymerizing telomeric DNA. Experimental telomerase upregulation in the mouse has been shown to confer proliferation properties on hair follicle stem cells but not on their progeny cells. Such experimental over-expression of the telomerase protein TERT, even in mice genetically deleted for the RNA component of telomerase (which therefore lack any telomerase enzymatic activity), specifically causes these stem cells to proliferate excessively (Sarin et al. 2005). This result showed that TERT can exert effects in vivo independent of its role in telomeric DNA polymerization. In cancer cells, which have high telomerase activity levels, partially knocking down telomerase RNA – by RNAi or ribozyme administration – rapidly caused the cells to change their properties, including gene expression profiles and morphology, even though they continued to divide (Li et al. 2005; Bagheri et al. 2006). Yeast and mammalian cells can maintain telomeres and quite successful cell growth rates even when telomerase is genetically deleted, through recombination-based pathways that, in essence, patch together telomeric tracts onto shortened telomeres through "borrowing' from other chromosomal telomeric tracts' DNA ends (Lundblad and Blackburn 1993). However, in yeast cells under such a telomerase-independent telomere maintenance regime, a sustained genome-wide expression response resembling an environmental stress response was observed, despite the fact that these cells seemed to be growing well (Nautiyal et al. 2002). There is also evidence for telomerase components in cells that are not dividing: TERT protein has been reported to be expressed in postmitotic hippocampal neurons even though telomerase enzymatic activity was not detected (Fu et al. 2000). Along with other hints (reviewed in Blackburn 2001, 2005), these findings point to possible functions for telomerase beyond its crucial and better-known function of maintaining telomere length in dividing cells. Thus the control of telomerase expression and activity is of great interest for all cells, including the stem cells of the brain and their dividing as well as postmitotic progeny.

Telomere Maintenance, Human Aging and Aging-Related Diseases

Research in the past two decades points to a link between organismal aging and aging-related diseases and cellular senescence caused by telomere shortening. Several lines of evidence strongly suggest that the resulting telomere dysfunction could have a causal role in some aging and aging-related diseases. White blood cells [leukocytes, or peripheral blood mononuclear cells (PBMCS)] are the most readily available source of normal human cells in which to measure telomere length or telomerase activity directly. Numerous clinical studies link short telomere length in white blood cells with aging-related disease or preclinical conditions of diseases. A short list of these conditions includes increased mortality from cardiovascular disease and infectious disease (Cawthon et al. 2003), heart disease (Starr et al. 2006; Brouilette et al. 2007) including coronary atherosclerosis (Samani et al. 2001), premature myocardial infarction and stroke (Brouilette et al. 2003; Fitzpatrick et al. 2007), vascular dementia (von Zglinicki et al. 2000), hypertension with carotid atherosclerosis (Benetos et al. 2004), age-related calcific aortic stenosis (Kurz et al. 2004), increased pulse pressure (Jeanclos et al. 2000) and stress (Epel et al. 2004), obesity and smoking (Valdes et al. 2005), osteoarthritis (Zhai et al. 2006), Alzheimer's disease (Panossian et al. 2003; Zhang et al. 2003), and insulin resistance, a preclinical condition for diabetes (Gardner et al. 2005; Adaikalakoteswari et al. 2007).

Finally, the strongest evidence suggesting a direct role of telomerase and telomere maintenance in aging and aging-related diseases came from study of the form of a rare human genetic disease, dyskeratosis congenita, caused by haploinsufficiency of telomerase activity due to mutations in hTER (Dokal and Vulliamy 2003). Dyskerotosis congenita patients with hTER or hTERT mutations have shorter telomeres and lower telomerase activity (Marrone et al. 2005). Patients die of eventual failure of the hematopoietic system, supporting the idea that premature senescence of the hematopoietic cells is one of the underlying causes of mortality (Marrone et al. 2005).

Chronic Psychological Stress, Telomerase, Aging and Aging-Related Diseases

Interestingly, cardiovascular diseases, neurodegenerative disease and immune dysfunction are aging-related diseases and are also stress-related diseases. Numerous epidemiological studies have shown that chronic stress leads to a poor health profile and to increased rates of stress-related diseases, including diabetes, cardiovascular diseases, mental illness and dampened immune functions (Raikkonen et al. 1996; Sapolsky 1996; Biondi and Zannino 1997; Kendler et al. 1999; Charney and Manji 2004; McEwen 2004; Rosengren et al. 2004; Yusuf et al. 2004; Glaser and Kiecolt-Glaser 2005; Lupien et al. 2005; Das and O'Keefe 2006; Shors 2006). Since telomere length is affected by telomerase activity, we tested whether telomerase activity in PBMCs might be affected by quantifiable measures of chronic psychological stress (Epel et al. 2004). We discovered that chronic stress is associated with at least two markers of cellular aging: notably, shorter telomere length and lower telomerase. We have reported, for the carefully controlled cohort of apparently healthy women aged between 20 and 50, that the number of years of chronic life stress, as well as perception of life stress, is related to lower telomerase activity and excessive telomere shortness in white blood cells. In the same cohort of women, shorter telomere length was also related to greater excretion of stress hormones (epinephrine, norepinephrine and cortisol) and lower telomerase was

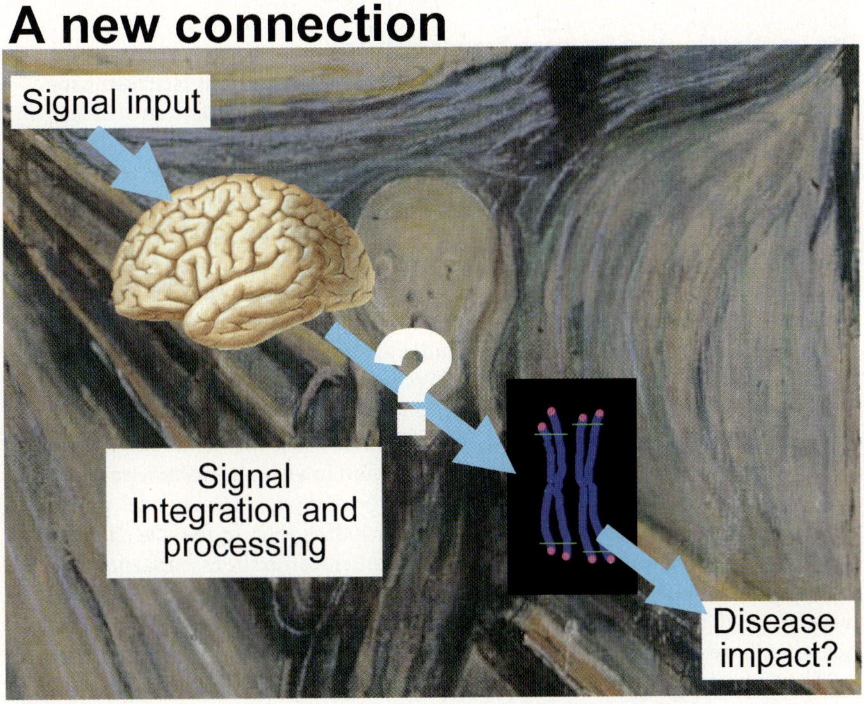

Fig. 3. A new connection between psychological stress, telomerase activity and human disease

related to more epinephrine excretion, over a 12-hour night time period (Epel et al. 2006). These findings suggested that stress arousal might be one of the mediators in the relation between psychological stress and cellular aging (Epel et al. 2006). Previous animal studies have shown that telomerase can also play a role in cardiovascular disease pathobiology, but the relationship had not been examined in humans until now. We found that women who had lower telomerase activity also had higher levels of risk for cardiovascular disease, as represented by a cluster of symptoms called the Metabolic Syndrome. Specifically, low telomerase (but not, in this relatively young cohort of women, telomere length) was associated with greater abdominal adiposity and higher blood pressure, cholesterol and blood sugars (Epel et al. 2006). These findings suggested for the first time that low telomerase in white blood cells may serve as a proxy of disease risk, possibly before telomere shortening occurs. We also found that women with low white blood cell telomerase (below the mean) responded to a standardized laboratory stressor with a decrease in vagal tone (heart rate variability). (Epel et al. 2006). This type of decrease is generally an indicator of less healthy cardiac function. Such responses to laboratory stress tend to have some traitlike characteristics (i.e., stability over time). Thus, we infer that habitually responding to stressful situations with this more malignant cardiovascular reactivity profile is linked to lower white blood cell telomerase. This work uncovered provocative new links between psychological stress arousal, impaired telomere maintenance and risk of heart disease.

The Implications of Reverse Transcription – Telomerase and Retrotransposition – in an Individual Human Life

Telomerase has evolved into an indispensable enzyme for the continued division of eukaryotic cells, and hence it plays an essential role in eukaryotic life cycles, including in every human life span. This fact is strikingly and starkly illustrated by the haplo-insufficiency for telomerase in humans described above: individuals with a mutation of the telomerase RNA gene that renders that allele non-functional die (apparently from exhaustion of stem cells or progenitor cells) before they can reach old age, even though their other telomerase RNA gene allele encodes a functional copy of the gene. Although McClintock proposed some decades ago that movement of mobile elements might be harnessed for developmental purposes (Fedoroff and Botstein 1992), until recently there has been little clear evidence for whether any transposons, including retroelements, might play a required role within one organism's lifetime. Hence, in contrast to telomerase's reverse transcription action in vivo, the reverse transcription associated with movement of retroelements had been thought to play roles that would manifest only over evolutionary time frames, including roles in the diversification of genomes and gene families (Yohn et al. 2005). In other words, the essential nature of the reverse transcriptase action of telomerase throughout life had been thought to distinguish it from the reverse transcription events mediated by other reverse transcriptases, including those of retroelements.

The discovery by Gage and collaborators (Muotri et al. 2005; Muotri and Gage 2006) that certain neuronal stem cell progeny (neural progenitor) cells undergo cell-type specific retroelement mobilization refocuses interest on the potential of retrotransposition for playing roles in any one human life. In the brain of each individual organism, the genomic alterations resulting from retroelement movements have the potential for a range of slightly differing genome readouts, not only in these somatic neural cells themselves but also in their cell division offspring. This discovery opens up the possibility that, in an individual's brain, function may be influenced by its unique history of retroelement movement events.

While retroelements are activated specifically in specific brain cells in the mouse, they do not apparently move actively in cells in general. Thus both telomerase and retroelement transcription have in common the feature that they are kept under tight downregulation control in mammalian cells. It will be of interest to see whether any transcriptional or other expression controls are shared between the telomerase reverse transcriptase and the reverse transcriptase of the retroelements mobilized in mouse brain stem cells. Human and mouse telomerase RNA and TERT are each regulated, at the transcriptional and post-transcriptional levels, by positive and negative control pathways (although the transcriptional control varies somewhat between these two species). Mammalian retroelement transcription is also controlled by a multiplicity of cell- and developmental stage-specific factors (for example, see Yu et al. 2001; Yang et al. 2003; Lavie et al. 2004; Xu and Blackburn 2004; Muckenfuss et al. 2006). Inspection of the known transcriptional control factors for human telomerase does not yet suggest any elements in common with those for the retroelements. However, as the control of each type of RT is complex and not fully worked out to date, there exists the possibility of shared controls that could be relevant for brain stem cell progeny functions. Further investigation needs to be done to follow up the provocative hint that a feature common to both these reverse transcriptases may be activation in stem cells or their immediate progeny.

References

Adaikalakoteswari A, Balasubramanyam M, Ravikumar R, Deepa R, Mohan V (2007) Association of telomere shortening with impaired glucose tolerance and diabetic macroangiopathy. Atherosclerosis 195:83–89. doi:10.1016/j.atherosclerosis.2006.12.003

Aisner DL, Wright WE, Shay JW (2002) Telomerase regulation: not just flipping the switch. Curr Opin Genet Dev 12:80–85.

Bagheri S. Nosrati M, Li S, Fong S, Torabian S, Rangel J, Moore DH, Federman S, Laposa RR, Baehner FL, Sagebiel RW, Cleaver JE, Haqq C, Debs RJ, Blackburn EH, Kashani-Sabet M (2006) Genes and pathways downstream of telomerase in melanoma metastasis. Proc Natl Acad Sci USA 103:11306–11311.

Benetos A, Gardner JP, Zureik M, Labat C, Xiaobin L, Adamopoulos C, Temmar M, Bean KE, Thomas F, Aviv A (2004) Short telomeres are associated with increased carotid atherosclerosis in hypertensive subjects. Hypertension 43:182–185. Epub 2004 Jan 2019.

Biondi M, Zannino LG (1997) Psychological stress, neuroimmunomodulation, and susceptibility to infectious diseases in animals and man: a review. Psychother Psychosom 66:3–26.

Blackburn EH (1993) Telomerase. In: Gesteland RF, Atkins JF (eds) The RNA world. Cold Spring Harbor Laboratory Press, Cold Spring Harbor, NY. 609–635

Blackburn EH (1999) Telomerase. In: Gesteland RF, Atkins JF (eds) The RNA world. Cold Spring Harbor Laboratory Press, Cold Spring Harbor, NY, pp. 609–635.

Blackburn EH (2001) Switching and signaling at the telomere. Cell 106:661–673.

Blackburn EH (2005) Cell biology: Shaggy mouse tales. Nature 436:922–923.

Brouilette S, Singh RK, Thompson JR, Goodall AH, Samani NJ (2003) White cell telomere length and risk of premature myocardial infarction. Arterioscler Thromb Vasc Biol 23:842–846. Epub 2003 Mar 2020.

Brouilette SW, Moore JS, McMahon AD, Thompson JR, Ford I, Shepherd J, Packard CJ, Samani NJ (2007) Telomere length, risk of coronary heart disease, and statin treatment in the West of Scotland Primary Prevention Study: a nested case-control study. Lancet 369:107–114.

Cawthon RM, Smith KR, O'Brien E, Sivatchenko A, Kerber RA (2003) Association between telomere length in blood and mortality in people aged 60 years or older. Lancet 361:393–395.

Charney DS, Manji HK (2004) Life stress, genes, and depression: multiple pathways lead to increased risk and new opportunities for intervention. Sci STKE 2004(225):re5.

Cong YS, Wright WE, Shay JW (2002) Human telomerase and its regulation. Microbiol Mol Biol Rev 66:407–425, table of contents.

Counter CM, Meyerson M, Eaton EN, Weinberg RA (1997) The catalytic subunit of yeast telomerase. Proc Natl Acad Sci USA 94:9202–9207.

Das S, O'Keefe JH (2006) Behavioral cardiology: recognizing and addressing the profound impact of psychosocial stress on cardiovascular health. Curr Atheroscler Rep 8:111–118.

Dokal I, Vulliamy T (2003) Dyskeratosis congenita: its link to telomerase and aplastic anaemia. Blood Rev 17:217–225.

Endoh T, Tsuji N, Asanuma K, Yagihashi A, Watanabe N (2005) Survivin enhances telomerase activity via up-regulation of specificity protein 1- and c-Myc-mediated human telomerase reverse transcriptase gene transcription. Exp Cell Res 305:300–311.

Epel ES, Blackburn EH, Lin J, Dhabhar FS, Adler NE, Morrow JD, Cawthon RM (2004) Accelerated telomere shortening in response to life stress. Proc Natl Acad Sci USA 101:17312–17315.

Epel ES, Lin J, Wilhelm FH, Wolkowitz OM, Cawthon R, Adler NE, Dolbier C, Mendes WB, Blackburn EH (2006) Cell aging in relation to stress arousal and cardiovascular disease risk factors. Psychoneuroendocrinology 31: 277–287.

Fedoroff NV, Botstein D (1992) The dynamic genome: Barbara McClintock's ideas in the century of genetics. Cold Spring Harbor Laboratory Press, Cold Spring Harbor, NY.

Feng J, Funk WD, Wang SS, Weinrich SL, Avilion AA, Chiu CP, Adams RR, Chang E, Allsopp RC, Yu J, Le S, West MD, Harley CB, Andrews WH, greider CW, Villeponteau B (1995) The RNA component of human telomerase. Science 269:1236–1241.

Fitzpatrick AL, Kronmal RA, Gardner JP, Psaty BM, Jenny NS, Tracy RP, Walston J, Kimura M. Aviv A (2007) Leukocyte telomere length and cardiovascular disease in the cardiovascular health study. Am J Epidemiol 165:14–21.

Forsyth NR, Wright WE, Shay JW (2002) Telomerase and differentiation in multicellular organisms: turn it off, turn it on, and turn it off again. Differentiation 69:188–197.

Fu W, Killen M, Culmsee C, Dhar S, Pandita TK, Mattson MP (2000) The catalytic subunit of telomerase is expressed in developing brain neurons and serves a cell survival-promoting function. J Mol Neurosci 14:3–15.

Gardner JP, Li S, Srinivasan SR, Chen W, Kimura M, Lu X, Berenson GS, Aviv A (2005) Rise in insulin resistance is associated with escalated telomere attrition. Circulation 111:2171–2177.

Glaser R, Kiecolt-Glaser JK (2005) Stress-induced immune dysfunction: implications for health. Nature Rev Immunol 5:243–251.

Greider CW, Blackburn EH (1985) Identification of a specific telomere terminal transferase activity in Tetrahymena extracts. Cell 43(2 Pt 1):405–413.

Greider CW, Blackburn EH (1987) The telomere terminal transferase of Tetrahymena is a ribonucleoprotein enzyme with two kinds of primer specificity. Cell 51:887–898.

Grieder CW, Blackburn EH (1989) A telomeric sequence in the RNA of Tetrahymena telomerase required for telomere repeat synthesis. Nature 337:331–337.

Harley CB, Futcher AB, Greider CW (1990) Telomeres shorten during ageing of human fibroblasts. Nature 345:458–460.

Jeanclos E, Schork NJ, Kyvik KO, Kimura M, Skurnick JH, Aviv A (2000) Telomere length inversely correlates with pulse pressure and is highly familial. Hypertension 36:195–200.

Kendler KS, Karkowski LM, Prescott CA (1999) Causal relationship between stressful life events and the onset of major depression. Am J Psychiat 156:837–841.

Koonin EV (2006) The origin of introns and their role in eukaryogenesis: a compromise solution to the introns-early versus introns-late debate? Biol Direct 1:22.

Kulpa DA, Moran JV (2006) Cis-preferential LINE-1 reverse transcriptase activity in ribonucleoprotein particles. Nature Struct Mol Biol 13:655–660.

Kurz DJ, Decary S, Hong Y,, Trivier E, Akhmedov A, Erusalimsky JD (2004) Chronic oxidative stress compromises telomere integrity and accelerates the onset of senescence in human endothelial cells. J Cell Sci 117:2417–2426.

Lavie L, Maldener E, Brouha B, Meese EU, Mayer J (2004) The human L1 promoter: variable transcription initiation sites and a major impact of upstream flanking sequence on promoter activity. Genome Res 14:2253–2260.

Lee MS, Blackburn EH (1993) Sequence-specific DNA primer effects on telomerase polymerization activity. Mol Cell Biol 13:6586–6599.

Lee MS, Gallagher RC, Bradley J, Blackburn EH (1993) In vivo and in vitro studies of telomeres and telomerase. Cold Spring Harb Symp Quant Biol 58:707–718.

Li S, Crothers J, Haqq CM, Blackburn EH (2005) Cellular and gene expression responses involved in the rapid growth inhibition of human cancer cells by RNA interference-mediated depletion of telomerase RNA. J Biol Chem 280:23709–23717.

Lin J, Ly H, Hussain A, Abraham M, Pearl S, Tzfati Y, Parslow TG, Blackburn EH (2004) A universal telomerase RNA core structure includes structured motifs required for binding the telomerase reverse transcriptase protein. Proc Natl Acad Sci USA 101:14713–14718.

Lingner J, Hughes TR, Shevchenko A, Mann M, Lundblad V, CechTR (1997) Reverse transcriptase motifs in the catalytic subunit of telomerase. Science 276:561–567.

Lundblad V, Blackburn EH (1993) An alternative pathway for yeast telomere maintenance rescues est1- senescence. Cell 73:347–360.

Lupien SJ, Fiocco A, Wan N, Maheu F, Lord C, Schramek T, Tu MT (2005) Stress hormones and human memory function across the lifespan. Psychoneuroendocrinology 30:225–242.

Malik HS, Eickbush TH, Goldfarb DS (1997) Evolutionary specialization of the nuclear targeting apparatus. Proc Natl Acad Sci USA 94:13738–13742.

Malik HS, Burke WD, Eickbush TH (2000) Putative telomerase catalytic subunits from Giardia lamblia and Caenorhabditis elegans. Gene 251:101–108.

Marrone A, Walne A, Dokal I (2005) Dyskeratosis congenita: telomerase, telomeres and anticipation. Curr Opin Genet Dev 15:249–257.

McEwen BS (2004) Protection and damage from acute and chronic stress: allostasis and allostatic overload and relevance to the pathophysiology of psychiatric disorders. Ann NY Acad Sci 1032:1–7.

Muckenfuss H, Hamdorf M, Held U, Perkovic M, Lower J, Cichutek K, Flory E, Schumann GG, Munk C (2006) APOBEC3 proteins inhibit human LINE-1 retrotransposition. J Biol Chem 281:22161–22172.

Muotri AR, Gage FH (2006) Generation of neuronal variability and complexity. Nature 441:1087–1093.

Muotri AR, Chu VT, Marchetto MC, Deng W, Moran JV, Gage FH (2005) Somatic mosaicism in neuronal precursor cells mediated by L1 retrotransposition. Nature 435:903–910.

Nakamura TM, Morin GB, Chapman KB, Weinrich SL, Andrews WH, Lingner J, Harley CB, Cech TR (1997) Telomerase catalytic subunit homologs from fission yeast and human. Science 277:955–959.

Nautiyal S, DeRisi JL, Blackburn EH (2002) The genome-wide expression response to telomerase deletion in Saccharomyces cerevisiae. Proc Natl Acad Sci USA 99:9316–9321.

Panossian LA, Porter VR, Valenzuela HF, Zhu X, Reback E, Masterman D, Cummings JL, Effros RB (2003) Telomere shortening in T cells correlates with Alzheimer's disease status. Neurobiol Aging 24:77–84.

Piskareva O, Schmatchenko (2006) DNA polymerization by the reverse transcriptase of the human L1 retrotransposon on its own template in vitro. FEBS Lett 580:661–668.

Raikkonen K, Keltikangas-Jarvinen L, Adlercreutz H, Hautanen A (1996) Psychosocial stress and the insulin resistance syndrome. Metabolism 45:1533–1538.

Ritz JM, Kuhle O, Riethdorf S, Sipos B, Deppert W, Englert C, Gunes C (2005) A novel transgenic mouse model reveals humanlike regulation of an 8-kbp human TERT gene promoter fragment in normal and tumor tissues. Cancer Res 65:1187–1196.

Rosengren A, Hawken S, Ounpuu S, Sliwa K, Zubaid M, Almahmeed WA, Blackett KN, Sitthi-amorn C, Sato H, Yusuf S (2004) Association of psychosocial risk factors with risk of acute myocardial infarction in 11119 cases and 13648 controls from 52 countries (the INTERHEART study): case-control study. Lancet 364:953–962.

Samani NJ, Boultby R, Butler R, Thompson JR, Goodall AH (2001) Telomere shortening in atherosclerosis. Lancet 358:472–473.

Sapolsky RM (1996) Why stress is bad for your brain. Science 273:749–750.

Sarin KY, Cheung P, Gilison D, Lee E, Tennen RI, Wang E, Artandi MK, Oro AE, Artandi SE (2005) Conditional telomerase induction causes proliferation of hair follicle stem cells. Nature 436:1048–1052.

Shors TJ (2006) Stressful experience and learning across the lifespan. Annu Rev Psychol 57:55–85.

Starr JM, McGurn B, Harris SE, Whalley LJ, Deary IJ, Shiels PG (2006) Association between telomere length and heart disease in a narrow age cohort of older people. Exp Gerontol 42:571–573

Valdes AM, Andrew T, Gardner JP, Kimura M, Oelsner E, Cherkas LF, Aviv A, Spector TD (2005) Obesity, cigarette smoking, and telomere length in women. Lancet 366:662–664.

von Zglinicki T, Pilger R, Sitte N (2000) Accumulation of single-strand breaks is the major cause of telomere shortening in human fibroblasts. Free Radic Biol Med 28:64–74.

Wright WE, Piatyszek MA, Rainey WE, Byrd W, Shay JW (1996) Telomerase activity in human germline and embryonic tissues and cells. Dev Genet 18:173–179.

Xu L, Blackburn EH (2004) Human Rif1 protein binds aberrant telomeres and aligns along anaphase midzone microtubules. J Cell Biol 167:819–830.

Yang N, Zhang L, Zhang Y, Kazazian HH Jr. (2003) An important role for RUNX3 in human L1 transcription and retrotransposition. Nucleic Acids Res 31:4929–4940.

Yohn CT, Jiang Z, McGrath SD, Hayden KE, Khaitovich P, Johnson ME, Eichler MY, McPherson JD, Zhao S, Paabo S, Eichler EE (2005) Lineage-specific expansions of retroviral insertions within the genomes of African great apes but not humans and orangutans. PLoS Biol 3(4):e110.

Yu F, Zingler N, Schumann G, Stratling WH (2001) Methyl-CpG-binding protein 2 represses LINE-1 expression and retrotransposition but not Alu transcription. Nucleic Acids Res 29:4493–4501.

Yusuf, S, Hawken S, Ounpuu S, Dans T, Avezum A, Lanas F, McQueen M, Budaj A, Pais P, Vari-gos J, Lisheng (2004) Effect of potentially modifiable risk factors associated with myocardial infarction in 52 countries (the INTERHEART study): case-control study. Lancet 364:937–952.

Zhai G, Aviv A, Hunter DJ, Hart, DJ, Gardner JP, Kimura M, LuX, Valdes AM, Spector TD (2006) Reduction of leucocyte telomere length in radiographic hand osteoarthritis: a population-based study. Ann Rheum Dis 65:1444–1448.

Zhang J, Kong Q, Zhang Z, Ge P, Ba D, He W (2003) Telomere dysfunction of lymphocytes in patients with Alzheimer disease. Cogn Behav Neurol 16:170–176.

Molecular and Circuit Mechanisms for Hippocampal Learning

Susumu Tonegawa[1] and *Thomas J. McHugh*[1]

The hippocampus is crucial for the formation of memories of facts and episodes (Scoville and Milner 1957; Jarrard 1993; Squire et al. 2004; Burgess et al. 2002). In storing the contents of a specific episode, the hippocampus must rapidly form and maintain representations of the temporal and spatial relationship of events and keep these representations distinct, allowing similar episodes to be distinguished, a property termed pattern separation. Furthermore, because specific episodes are rarely replicated in full, the hippocampus must be capable of using partial cues to retrieve previously stored patterns of representations, a phenomenon referred to as pattern completion. Based primarily on the anatomy (Fig. 1) and physiology of the hippocampus and its associated cortical structures, computational neuroscientists have suggested specific hippocampal subregions and circuits that may subserve these mnemonic requirements. These are the feedforward pathway from the entorhinal cortex (EC) to the dentate gyrus (DG) and on to CA3 for pattern separation, and the recurrent and highly plastic connections in CA3 for pattern completion (Marr 1971; McClelland and Goddard 1996; McNaughton and Nadel 1990; O'Reilly and McClelland 1994).

CA3 NMDA Receptors for Pattern Completion

CA3 pyramidal cells receive excitatory inputs from three sources: the mossy fibers of the DG granule cells (GC), the perforant path axons of the stellate cells in the superficial layers of the EC, and the recurrent collaterals (RC) of the CA3 pyramidal cells and, in return, provide output to CA1 pyramidal cells via Schaffer Collaterals (SC). The prominence of these RCs has led to suggestions that CA3 might engage these connections to serve as an associative memory network. Associative networks, in which memories are stored through modification of synaptic strength within the network, are capable of retrieving entire memory patterns from partial or degraded inputs (pattern completion; Marr 1971; Gardner-Medwin 1976; Hopfield 1982; McNaughton and Morris 1987; Rolls 1989; Hasselmo et al. 1995).

We set out to obtain evidence for this hypothesis by targeting the knockout of the NR1 gene, coding for the essential subunit of NMDA receptors, to postnatal CA3 pyramidal cells. Use of the Cre-loxP recombination system, in which the expression

---

[1] The Picower Institute for Learning & Memory, RIKEN-MIT Neuroscience Research Center, Massachusetts Institute of Technology, 77 Massachusetts Avenue, Cambridge MA 02139-4307, USA, e-mail: tonegawa@mit.edu

Gage et al.
Retrotransposition, Diversity and the Brain
© Springer-Verlag Berlin Heidelberg 2008

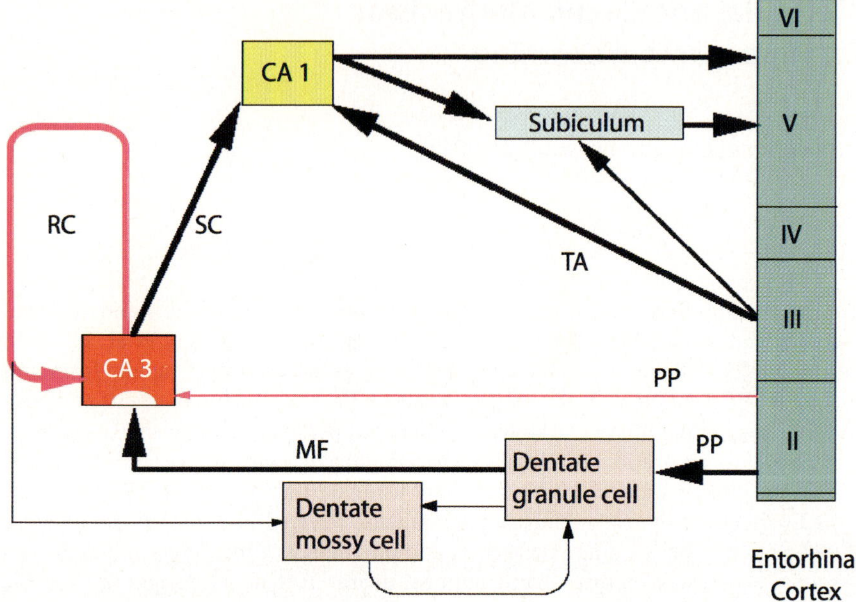

Fig. 1. Hippocampal excitatory pathway. PP = Perforant Path; MF = Mossy Fiber; RC = Recurrent Collaterals; SC = Schaffer Collaterals; TA = Temporoammonic Path. The feedforward pathway from entorhinal cortext to dentate gyrus and to CA3 is hypothesized to play a role in pattern separation while the RC in CA3 is hypothesized to be important for pattern completion

of the transgenic Cre gene was driven by the transcription-regulating elements of the KA-1 gene, permitted us to obtain such a cell type-specific knockout mouse (CA3-NR1 KO; Nakazawa et al. 2002). In situ hybridization, immunohistochemistry, and hippocampal slice electrophysiology data confirmed that the knockout is restricted to postnatal CA3 pyramidal cells.

The CA3-NR1 KO mice, in contrast to CA1-NR1 KO mice (Tsien et al. 1996; McHugh et al. 1996), were normal in the standard hidden platform version of the Morris water maze task (Fig. 2). However, when the probe test was conducted under the conditions where only one of the four major visual cues (partial cue condition) was available after the training was performed with the four cues (full cue condition), the mutant mice exhibited a deficit compared to the control littermates (Fig. 2). These data indicate that the mutant mice are capable of acquiring this spatial memory and also retrieving it as long as the full set of cues are provided during the recall phase. However, the mutants are impaired in retrieving the memory using a partial set of cues (only one of the four major visual cues), conditions that are sufficient for recall in the control mice. These data suggest that the CA3-NR1 mice suffer from a specific impairment in a pattern completion-mediated recall.

This phenotype of the mutants observed at the behavioral level was corroborated at the level of neuronal ensemble activities in the CA1 area, which was shown by in vivo recording of CA1 pyramidal cells with the tetrode recording technique. The CA3-NR1 KO mice exhibited compact place fields that were indistinguishable from those of the

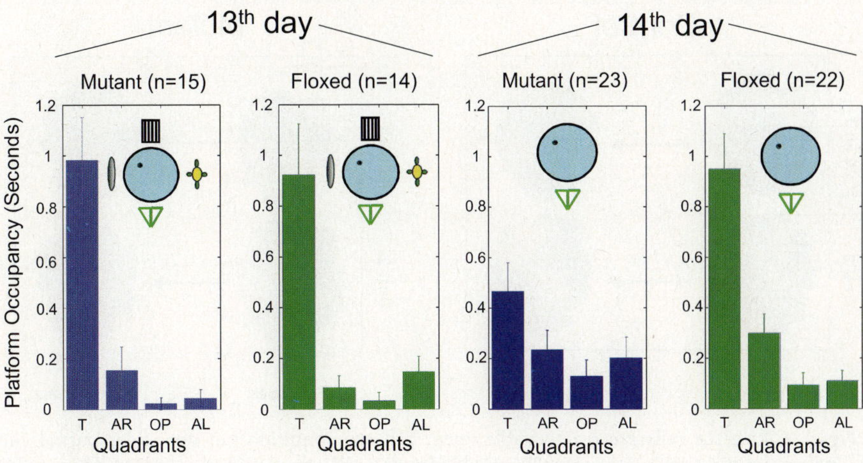

Fig. 2. The CA3-NR1 KO mice are defective in pattern completion. The mutants and control littermates (floxed) went through 12 day-long training in the hidden platform version of the Morris watermaze task under the full cue conditions (four visual cues surrounding the pool). The probe trial conducted on the 13th day under the same full cue condition indicated the mutant is normal in the acquisition and retrieval of the spatial memory under these conditions. However, the memory retrieval by the mutants was substantially diminished compared to the control mice in the probe trial conducted the next day (14th day) under the partial cue condition (only one of the four cues was available during this probe trial)

control mice under the familiar full cue conditions (four major visual cues; Fig. 3). When these mice were transferred to a home cage and after several hours returned to the same recording box with the full set of cues, the mutant place cells were reactivated as well as the control place cells. However, when the mice were returned to the recording box with a partial set of cues (only one of the four major visual cues), the extent of the reactivation of the place cells by the mutants was significantly diminished compared to the control mice (Fig. 3).

Thus, both behavioral and in vivo physiological data strongly support the hypothesis that the NMDA receptors in the CA3 pyramidal cells, and probably synaptic plasticity at the CA3-CA3 recurrent synapses, play a crucial role in pattern completion in the hippocampus.

DG NMDA Receptors for Rapid Pattern Separation

The key data that support the hypothesis that the feedforward EC $\rightarrow$ DG $\rightarrow$ CA3 pathway may be responsible in pattern separation are that 1) the number of DG GCs is substantially greater than the numbers of EC layer II stellate cells and CA3 pyramidal cells, 2) the connection between DG and CA3 is two orders of magnitude more sparse than the connections between other regions, including EC and DG, and 3) the DG GC spiking activity is lower compared to other regions. It is therefore possible that relatively overlapping memory engrams present in EC are separated (orthogonalized)

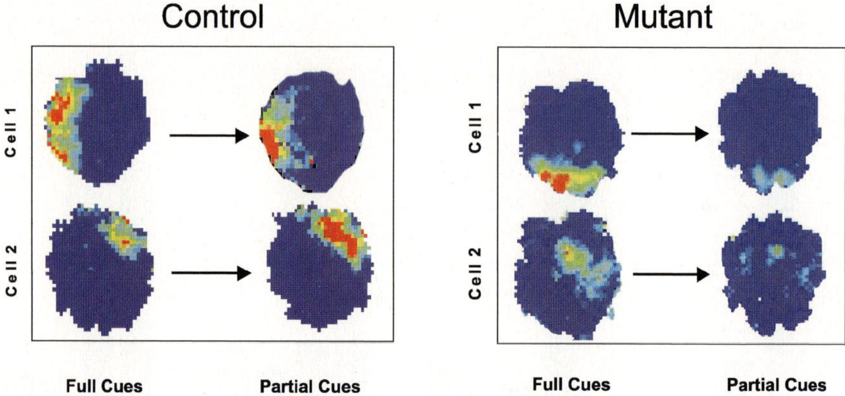

Fig. 3. CA1 place cells are reactivatable under full cue condition but not under partial cue condition in CA3-NR1 KO mice. The CA3-NR1 KO mice formed compact CA1 place fields in a familiar environment under full cue conditions. Upon a reexposure these place cells were reactivated well under full cue conditions (four cues), but only poorly under partial cue conditions (one cue)

as the information is transferred through the EC → DG → CA3 pathway. Since NMDA receptors in DG GCs are expected to modulate the activity of DG GCs in an experience-dependent manner, it is possible that we may see a deficit in an experience-dependent pattern separation in mutant mice in which the NR1 gene knockout is targeted to postnatal GCs.

We generated such NR1 knockout mice (DG-NR1 KO mice) fortuitously by employing the transcriptional regulatory elements of the proopiomelanocortin (POMC) gene as the driver of the Cre expression and crossing the Cre transgenic mice with the same "floxed" NR1 gene mice that we previously used for the CA3 (Nakazawa et al. 2002) and CA1 (Tsien et al. 1996; McHugh et al. 1996) studies (McHugh et al. 2007). Again, in situ hybridization in these mice, immunohistochemistry, and synaptic electrophysiology confirmed that the NR1 knockout is well restricted to postnatal DG GCs.

The performance of the DG-NR1 KO mice was normal in the standard Morris water maze task as well as in the standard contextual fear conditioning. However, in an incremental context discrimination fear conditioning task, the mutant mice exhibited a deficit in the early phase of the trials, although their ability to discriminate the contexts developed slowly to the normal level as the trials were repeated. Thus, the mutant mice were normal in spatial and contextual learning per se but had a problem in being able to rapidly distinguish similar contexts with just a few trials, which the control littermates accomplished with no problem. These results suggest that the NMDA receptors in DG GCs and probably NMDA receptor-dependent synaptic plasticity at the perforant path-DG GC synapses play an important role in fast (with one or two trials) pattern separation. However, the fact that the mutant mice can catch up to the control mice with more trials suggests that the hardwiring in the EC → DG → CA3 pathways permits slow, multitrial-dependent acquisition of pattern separation.

To detect a pattern separation deficit of the DG-NR1 KO mice at the neuronal ensemble activity level, we recorded with the tetrode technique the spiking activities in

CA1 and CA3 as the mice explored two distinct contexts (low-walled white circular box vs. black square box) at the same site in the same room. Earlier studies with normal rats had shown that, under these conditions, individual pyramidal cells in CA1 exhibited similar firing rates in the two contexts whereas those in CA3 displayed context-specific firing rates (Leutgreb et al. 2005). Thus, in the latter case, there was a "remapping" of the firing rates as the animals were shifted from one context to the other. Like rats, our control mice showed significant rate remapping in CA3 but no remapping in CA1. In contrast, the DG-NR1 KO mice exhibited a significant deficit in rate remapping in CA3 but no remapping in CA1. These results corroborate the behavioral deficit of contextual discrimination and reinforce the conclusion that DG NMDA receptors play a role in rapid pattern separation.

CA1 for Novelty Detection?

Our earlier study, carried out by applying the same interdisciplinary strategy to CA1 pyramidal cells, demonstrated that a knockout of NMDA receptors in this "outpost" of the excitatory hippocampal trisynaptic pathway leads to a severe impairment in hippocampus-dependent learning tests, such as the Morris water maze and trace fear conditioning (Tsien et al. 1996; McHugh et al. 1996). This finding is in contrast to the knockout of the same NR1 gene in CA3 or DG, but it is not surprising because, in the CA1-NR1 KO mice, the NMDA receptors are knocked out not only at the SC-CA1 synapses, the most downstream of the trisynaptic pathway, but also at the temporoammonic (TA) path-CA1 synapses, an integral part of the direct EC→CA1→(subiculum)→EC pathway. There has been a suggestion that inputs from these two circuits (trisynaptic and temporoammonic) are "compared" at CA1 to generate a "novelty signal" that may be necessary to convert the hippocampus to a "learning mode" (Fig. 4; Vinogradova 2001; Lisman 2005). After all, we may learn something when we encounter novelty whereas we cannot learn from something we already know. To address this postulated function of CA1, we need a new genetic manipulation technique that will allow us to block the SC input to CA1 specifically while keeping the TA input intact or vice versa. Such a technique is under development.

The genetic technology that permits a cell-type specific and postnatal knockout of a gene (such as the NR1 gene) and multidisciplinary analyses of these conditional mutant mice are allowing us to test a number of hypotheses regarding the distinct functions of hippocampal subregions and their circuits in various aspects of hippocampus-dependent learning and memory. In the future, this general strategy could be extended to brain systems and circuits outside of the hippocampus to uncover mechanisms underlying memory and other cognitive functions.

Acknowledgements. We wish to thank many people who participated in the work outlined in this short monograph. Many of them are co-authors of the original research papers from our laboratory. Special thanks go to Kazu Nakazawa, Michael Fanselow, Matt Jones, Matt Wilson and Nina Balthasar. The work was supported by NIH grants P50-MH 58880 and RO1-MH78821.

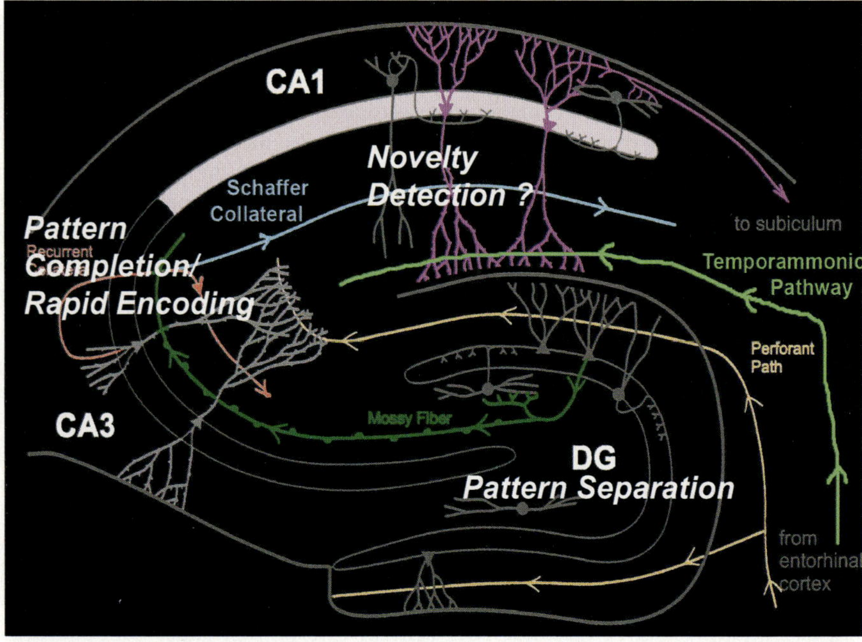

Fig. 4. Distinct mnemonic functions of hippocampal excitatory circuits. The cell type-restricted knockout technology revealed that the NMDA receptors in CA3 pyramidal cells and DG granule cells are important for pattern completion and pattern separation, respectively. The CA3 NMDA receptors also play a role in rapid encoding of one trial/experience memory (Nakazawa et al. 2003). It is hypothesized that CA1 pyramidal cells may compare the SC input which may be loaded with previously acquired memory information and the TA input which conveys on-line sensory input and, thereby, provides novelty/familiarity signal to the downstream areas like subiculum. The novelty signal is thought to be important to convert the hippocampal to a "learning mode"

References

Burgess N, Maguir EA, O'Keefe J (2002) The human hippocampus and spatial and episodic memory. Neuron 35:625–641

Gardner-Medwin AR (1976) The recall of events through the learning of associations between their parts. Proc Roy Soc London Ser B 194:375–402

Hasselmo ME, Schnell E, Barkai E (1995) Dynamics of learning and recall at excitatory recurrent synapses and cholinergic modulation in rat hippocampal region CA3. J Neurosci 15:5249–5262

Hopfield JJ (1982) Neural networks and physical systems with emergent collective computational abilities. Proc Natl Acad Sci USA 79:2554–2558

Jarrard LE (1993) (1993). On the role of the hippocampus in learning and memory in the rat. Behav Neural Biol 60:9–26

Leutgeb JK, Barnes CA, Moser EI, McNaughton BL, Moser MB (2005) Independent codes for spatial and episodic memory in hippocampal neuronal ensembles. Science 309:619–623

Lisman JE, Grace AA (2005)The hippocampal-VTA loop: controlling the entry of information into long-term memory. Neuron 46:703–713

Marr D (1971) (1971). Simple memory: a theory for archicortex. Phil Trans Roy Soc London 262:23–81

McClelland JL, Goddard NH (1996) Considerations arising from a complementary learning systems perspective on hippocampus and neocortex. Hippocampus 6:654–665

McHugh TJ, Blum KI, Tsien JZ, Tonegawa S, Wilson MA (1996) Impaired hippocampal representation of space in CA1-specific NMDAR1 knockout mice. Cell 87:1339–1349

McHugh TJ, Jones MW, Quinn JJ, Balthasar N, Coppari R, Elmquist JK, Lowell BB, Fanselow MS, Wilson MA, Tonegawa S (2007) Dentate Gyrus NMDA Receptors mediate rapid pattern separation in the hippocampal network. Science 317:94–9

McNaughton BL, Morris RGM (1987) Hippocampal synaptic enhancement and information storage within a distributed memory system. Trends Neurosci 10:408–415

McNaughton BL, Nadel L (1990) Hebb-Marr networks and the neurobiological representation of action in space. In: Gluck MA, Rumelhart D (eds) Neuroscience and connectionist theory. Erlbaum, Hillsdale, NJ, pp. 1–63

Nakazawa K, Quirk MC, Chitwood RA, Watanabe M, Yeckel MF, Sun LD, Kato A, Carr CA, Johnston D, Wilson MA, Tonegawa S (2002) Requirement for hippocampal CA3 NMDA receptors in associative memory recall. Science 297:211–218

Nakazawa K, Sun LD, Quirk MC, Rondi-Reig L, Wilson MA, Tonegawa S (2003) Hippocampal CA3 NMDA receptors are crucial for memory acquisition of one-time experience. Neuron 38:305–315

O'Reilly RC, McClelland JL (1994) Hippocampal conjunctive encoding, storage, and recall: avoiding a trade-off. Hippocampus 4:661–682

Rolls ET (1989) The representation and storage of information in neuronal networks in the primate cerebral cortex and hippocampus. In: Burbin R, Miall C, Mitchison G (eds) The computing neuron. Addison-Wesley, Wokingham, UK, pp. 125–159

Scoville WB, Milner B (1957) Loss of recent memory after bilateral hippocampal lesions. J Neuropsychiat Clin Neurosci 12:103–113

Squire LR, Stark CE, Clark RE (2004) The medial temporal lobe. Annu Rev Neurosci 27:279–306

Tsien JZ, Huerta PT, Tonegawa S (1996) The essential role of hippocampal CA1 NMDA receptor-dependent synaptic. Cell 87:1327–1338

Vinogradova OS (2001) Hippocampus as comparator: role of the two input and two output systems of the hippocampus in selection and registration of information. Hippocampus 11:578–598

Retrotransposons – Natural and Synthetic

Jef D. Boeke[1], *Wenfeng An*[1], *Lixin Dai*[1], *Edward S. Davis*[1], *Jeffrey S. Han*[1], *Kathryn A. O'Donnell*[1], *Lisa Z. Scheifele*[1], and *Sarah J. Wheelan*[1]

Transposable elements are ubiquitous among sequenced genomes. The host genomes roughly subdivide into two types: 1) streamlined, that is, small, with little space between genes and lacking large introns, or 2) bulky, with lots of space between genes and many large introns. Most microorganisms, along with selected vertebrates like the pufferfish, fall into the first class, whereas mammals and most plants fall into the second class. As can be seen from Fig. 1, transposable element abundance mirrors the genome type of the host, with mobile elements comprising half or more of many of these bulky genomes! Mobile elements are of two basic types: DNA transposons, which predominantly mobilize via a cut and paste mechanism, and retrotransposons, which move by a copy and paste mechanism involving reverse transcription of an RNA intermediate (Fig. 1 right panel; Curcio and Derbyshire 2003). Retrotransposons are found in virtually all eukaryotes, from yeast (Kim et al. 1998) to human (Lander

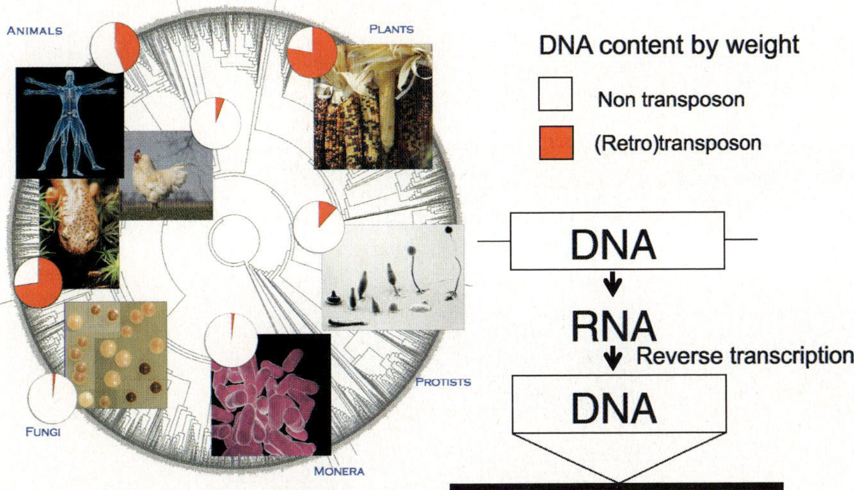

Fig. 1. *Left panel* shows the phylogenetic tree of life as determined by rDNA sequence alignments. Selected organisms are shown, along with the fraction of their genome made up of mobile elements as pie charts. On the *right* is the basic information flow used in the retrotransposition process

[1] High Throughput Biology Center, Johns Hopkins University School of Medicine, Baltimore, MD 21205, USA, e-mail: jboeke@jhmi.edu

Gage et al.
Retrotransposition, Diversity and the Brain
© Springer-Verlag Berlin Heidelberg 2008

et al. 2001). Remarkably, in a yeast cell, the number of retrotransposon copies can be changed rather dramatically without a major impact on the phenotype of the host. The change in copy number can be seen using a new tiling array technique by which it is possible to comprehensively map the unique genomic regions adjacent to all transposable element copies probed (Fig. 2; Wheelan et al. 2006). The ability of yeast strains to tolerate very high copy numbers of transposons is due in part to the fact that, in yeast, most insertions are targeted to non-essential genomic regions, even though most of the genome is protein-coding (Chalker and Sandmeyer 1990; Devine and Boeke 1996; Ji et al. 1993; Zou et al. 1996). This property and many others suggest that retrotransposons are highly coevolved with their hosts.

L1 retrotransposons or LINE-1s are ubiquitous mammalian mobile elements. Each mammalian species' genome is littered with copies of an L1 species that has coevolved with its genome (Gibbs et al. 2004; Kirkness et al. 2003; Lander et al. 2001; Waterston et al. 2002). L1 elements directly make up about 17% of our genome and are responsible

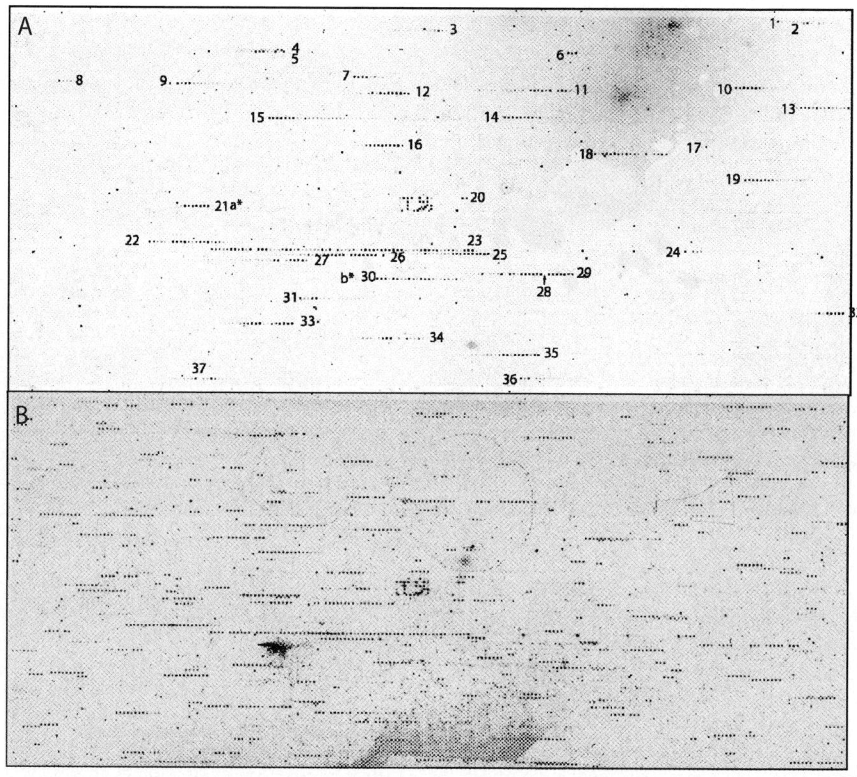

Fig. 2. Mapping transposon insertion sites. Genomic regions adjacent to transposons are PCR amplified and then identified by hybridization to a tiling array. Positive hybridization controls produce a visible "TY" signal. Because features on the array are ordered by chromosomal location, hybridization to adjacent features can be used to identify insertion sites in a wild-type yeast strain (**A**) or a strain with high transposon copy number (**B**)

for at least a third of our DNA by weight because they provide the molecular machinery for mobilizing not only their own sequences but also the highly abundant Alu sequences (Dewannieux et al. 2003), as well as the less abundant processed pseudogenes (Esnault et al. 2000). The latter "retrotranscripts" are simply cellular mRNAs that have been reverse transcribed by the L1 machinery and inserted into the genome, very much like L1 itself is inserted. Retrotransposons move in the genome via a replicative process (Fig. 3). After being transcribed into a full length RNA by host RNA polymerase, the RNA can be translated to produce two proteins, ORF1p and ORF2p. Together with the RNA, these form a ribonucleoprotein (RNP) complex (Martin 1991), which is imported into the nucleus. ORF2p has endonuclease (Feng et al. 1996) and reverse transcriptase (Mathias et al. 1991) functions essential for retrotransposition (Moran et al. 1996). The endonuclease selects and cleaves the target site (Cost and Boeke 1998), and the RNA is ultimately reverse transcribed to make a new retrotransposon copy, a process known as target-primed reverse transcription (TPRT; Fig. 4; Luan et al. 1993).

In somatic and tissue culture cells, L1 expression and hence transposition appear to be tightly regulated transcriptionally, and so the promoter that drives this expression

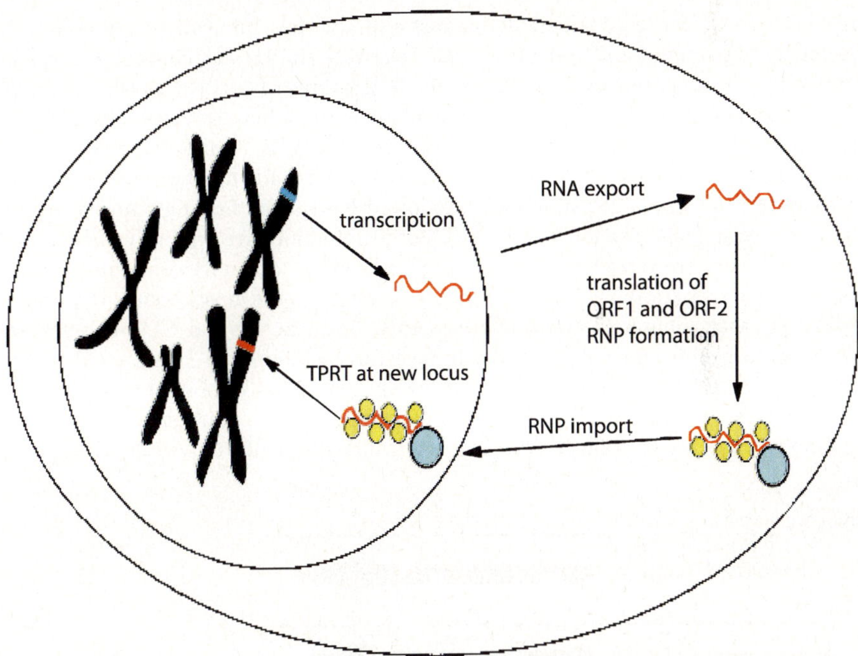

Fig. 3. Replicative cycle of L1 elements. A "donor" element (*blue band* on chromosome) is transcribed in the nucleus. The RNA (*red waved line*) is exported to the cytoplasm, where it is translated into ORF1 (*yellow spheres*) and ORF2 (*blue sphere*) proteins. The ribonucleoprotein (RNP) complex is imported into the nucleus and used as the machinery to drive target-primed reverse transcription (TPRT)/integration of a new copy of the element at a new locus (*red band* on chromosome)

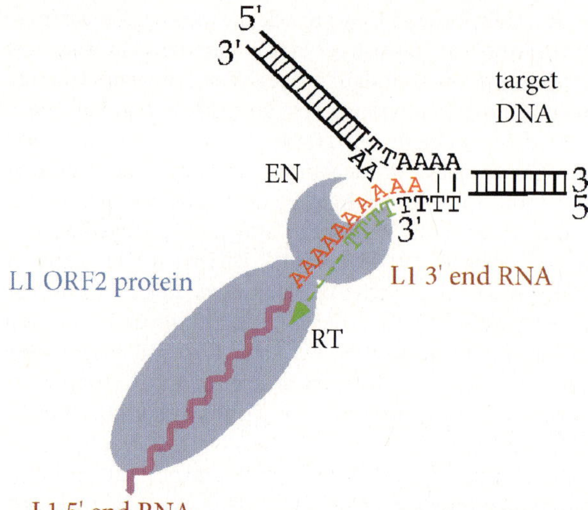

Fig. 4. Mechanism of LINE-1 integration by TPRT. The endonuclease (EN) domain of ORF2 creates a single-strand nick in the target DNA. The L1 RNA anneals with the DNA and ORF2's reverse transcriptase (RT) activity uses the target DNA's 3'-OH to prime synthesis of first strand cDNA

has been an object of considerable interest and scrutiny. Interestingly, although L1 elements in primates and rodents encode relatively similar proteins (percentage of amino acid identity ranges from 20% at the N-terminus of ORF1 to >60% in ORF2), the promoters not only lack sequence homology entirely but also have very different structures (Fig. 5). Most mouse L1 promoters (in the A, F, T^F and G^F subfamilies of mouse L1 elements), like those of other rodent L1s, are made up of a series of tandem repeats of ~200 bp, called monomers, followed by a short non-monomeric region (Goodier et al. 2001; Padgett et al. 1988). Both subfamilies are relatively ancient and most members are inactive. T^F is a young, expanding subfamily containing ~3 000 full-length members and ~1 800 of them are active. G^F is the most recently discovered subfamily that contains ~400 active elements. Both the T^F and G^F monomer are 70% identical to F-type monomer but differ from each other by 33%. In addition to the differences among monomer sequences, the numbers of monomer repeats and monomer lengths vary among individual element copies. In contrast, the human L1 promoter sequence in transpositionally and transcriptionally active (Ta) elements is

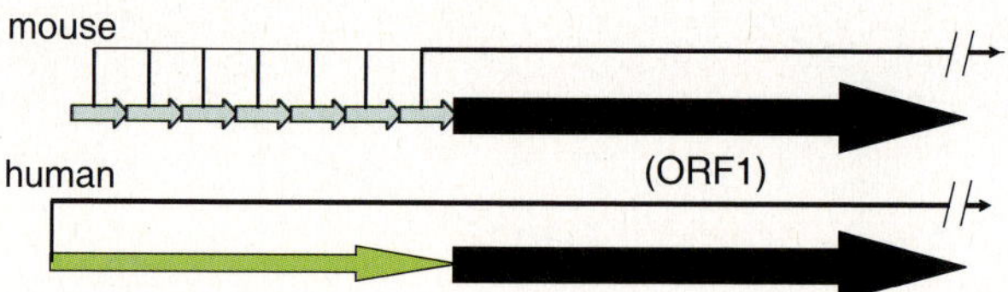

Fig. 5. Comparison of mouse and human L1 promoters. The 5' UTR region of most mouse L1 contains several tandem repeats (monomer) in the length of ~200 bp. Each *blue arrow* represents a monomer sequence. The 5' UTR of human L1 contains a ~900 bp, non-repetitive region (*yellow arrow*) that drives the transcription of L1 element. *Black arrow* denotes the first open reading frame of L1 (ORF1) and *fine line arrow* indicates the transcription direction

about 900 bp long, nonrepetitive and well-conserved in length, and it contains all of the elements required for transcription downstream of the transcription start site (Swergold 1990).

Selfish Gene?

In his book "The Selfish Gene," Richard Dawkins outlines the idea that evolution is driven at the level of individual genes. There is no more compelling example of this than mobile genetic elements like retrotransposons, to which host genomes/organisms are

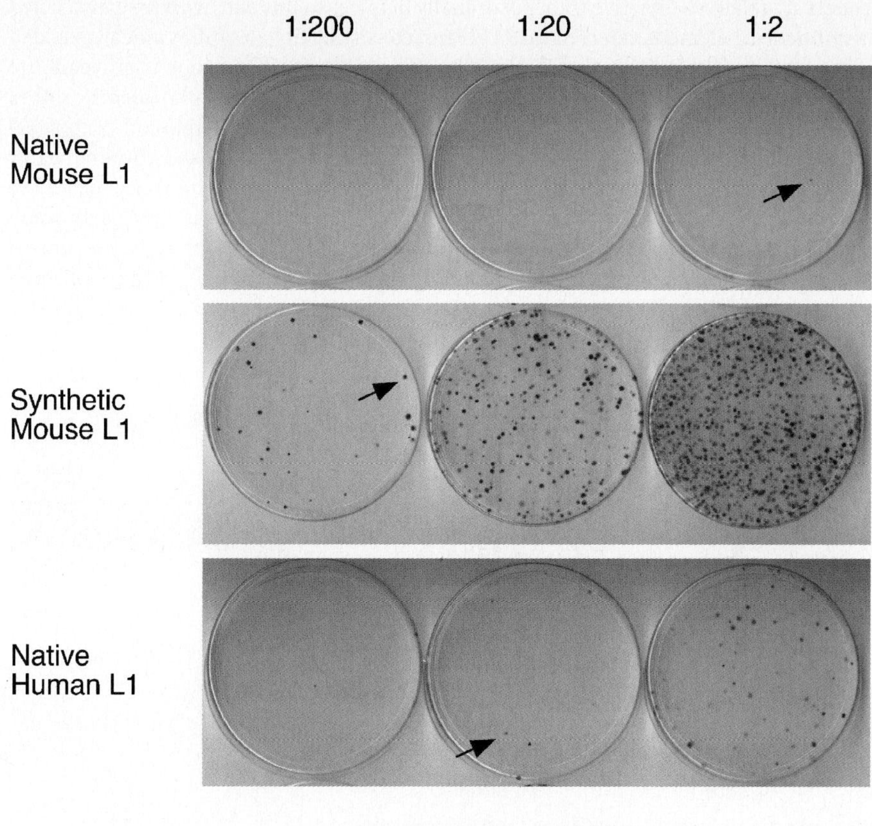

Fig. 6. Synthetic mouse L1 is much more active for retrotransposition than native mouse and human L1 elements. Retrotransposition assay was performed in HeLa cells for native mouse L1, synthetic mouse L1 and native human L1 elements, all of which were tagged with an intron-interrupted neomycin resistance gene reporter. L1 function is scored as the number of G418-resistant colonies because only when L1 completes one round of retrotransposition does a cell become G418-resistant. Cells were diluted at ratios as indicated prior to G418 selection

nothing more than "bags of genes" to exploit (Dawkins 1976). Like virtually all transposable elements found in metazoans, L1 element transposition was until relatively recently thought to be entirely germ-line specific, as predicted from strict "selfish gene" theory. However, recent findings indicate that L1s are highly active transcriptionally in mouse neuronal progenitor cells, and engineered human elements retrotranspose in mouse brain in a neuron-specific manner (Muotri et al. 2005).

Fully Synthetic Retrotransposons Are Highly Active

L1 retrotransposons are potential tools for in vivo mutagenesis; however, native L1 elements are relatively inactive transpositionally in mice. To this end, we have constructed a synthetic L1 element, referred to as *ORFeus*, consisting of two synonymously recoded open reading frames (Han and Boeke 2004). The sequence is based on a native mouse L1 element sequence, L1*spa* (Mulhardt et al. 1994) and can be controlled by either generic (e.g., CMV or CAG promoter) or native L1 5' end transcriptional control sequences. Such donor element constructs can be marked by a transposition indicator gene, which is inserted in the antisense orientation relative to the transcription of the *ORFeus* element. The reporter, either *neo* or *gfp*, is interrupted by an intron in the same sense as the *ORFeus* donor element. In this way, the donor element does not express the reporter because its coding region is disrupted by an inverse intron, but upon retrotransposition, the intron is removed during the RNA step and an active reporter gene is generated.

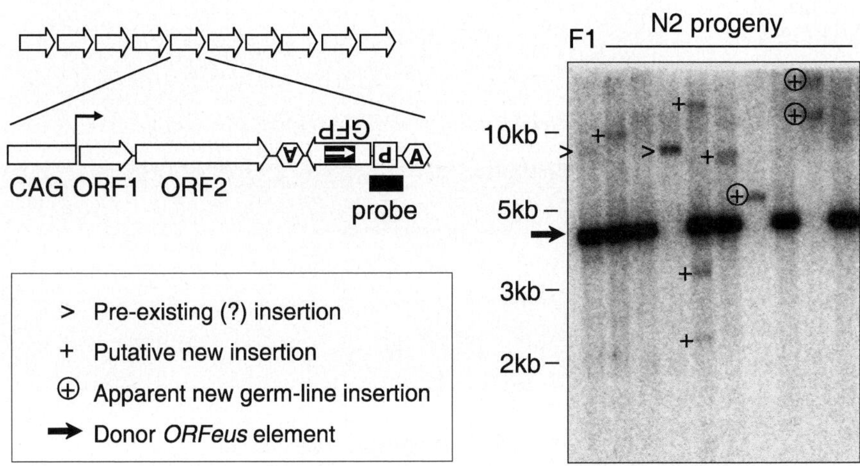

Fig. 7. Estimating germ-line insertion frequency by Southern blot analysis. The *top left panel* is a schematic of the 10-copy *ORFeus* donor transgene concatemer with a detailed view of the structure of a single-copy transgene under the regulation of CAG promoter and marked by an intron-disrupted GFP reporter cassette driven by its own promoter and polyadenylation site. The position of the Southern probe is indicated. The *right panel* is a Southern blot for nine N2 progeny mice from breeding their F1 transgenic parent (the first lane) to a wild type mouse

Using a *neo* intron removal assay, *ORFeus* was found to be ~200-fold more active for retrotransposition in cell culture than native mouse L1 elements and was even more active than the most active human elements studied previously (Fig. 6). To study *ORFeus* activity in vivo, we developed transgenic mouse models in which *ORFeus* expression was controlled by the constitutively active heterologous CAG promoter, and we measured *ORFeus* retrotransposition activity both in germ-line and somatic tissues (An et al. 2006). Germ-line retrotransposition frequencies resulting in 0.3–0.4 insertions per animal were seen among progeny of *ORFeus* donor element heterozygotes, as determined by Southern blotting (Fig. 7). This germ-line retrotransposition frequency compares favorably with previously observed retrotransposition frequencies with native elements driven by heterologous promoters (Babushok et al. 2006). Interestingly, we also observed somatic transposition events in 100% of these *ORFeus* donor-containing animals, and many different insertions were readily recovered from each animal using a modified inverse PCR protocol. Modeling exercises suggest that the numbers of somatic insertions per animal could be as high as millions, suggesting that these animals could provide important new models for cancer, as has recently been reported for the *Sleeping Beauty* DNA transposon (Collier et al. 2005; Dupuy et al. 2005). Somatic retrotransposition was observed in all tissues tested, including brain, but was not particularly elevated in any specific tissue in these mice driven by the CAG promoter. Nearly 200 insertions were precisely mapped, and their distribution in the mouse genome appeared random relative to transcription units and GC content (Fig. 8). Constitutive *ORFeus* may be extraordinarily useful for in vivo mouse mutagenesis. Gene traps are being developed for these purposes.

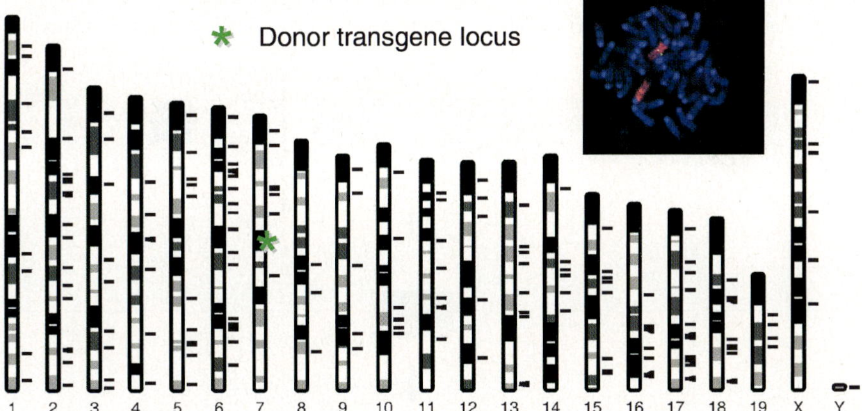

Fig. 8. Chromosomal distribution of mapped insertions. A total of 171 mappable insertions were charted to mouse genome build 36 (*short black lines* to the right of individual chromosomes). The approximate position of the donor concatemer on chromosome is marked (*green asterisk*), which was located by fluorescent in situ hybridization (shown in the insert). *Insert*: Metaphase spreads of splenocytes from donor-containing mice were probed with fluorescently labeled full-length transgene cDNA probe (*green*) and subsequently with a whole-chromosome paint probe for chromosome 7 (*red*). Chromosomes were counterstained with DAPI (*blue*)

Neural-specific Retrotransposition – Could it be Conserved from Rodents to Primates?

L1 elements in mouse and human, like most metazoan retrotransposons, show evidence for germ-line-specific expression (Branciforte and Martin 1994; Ergun et al. 2004; Trelogan and Martin 1995). There is evidence that native rodent L1s are active in neural progenitor cells stimulated to differentiate in response to FGF-2 and are upregulated transcriptionally. On this basis, Muotri et al. (2005) introduced a human L1 (driven by a human L1 promoter) marked with a retrotransposition indicator gene into such cells and into transgenic mice. Interestingly, when the cells were differentiated in tissue culture into astrocytes, glia and neurons, retrotransposition of the human constructs was only seen in those cells that differentiated into neuron-like cells. In some of these cases, insertion of the new retrotransposons was into transcriptionally active target genes in the differentiating neurons. Furthermore, significant retrotranspositional activity of this element (as inferred from GFP staining) was observed in a wide variety of neuronal cells in the brains of these mice. These results can be interpreted to suggest that the highly divergent promoters of primate and rodent LINEs, as well as the divergent proteins encoded by these elements, might be under genetic selection for retrotranspositional activity in the brain. Not only are these promoters highly divergent structurally, but there is also good reason to believe they have an independent genetic origin. The "promoter capture" model (Khan et al. 2006) posits that, as the host inactivates L1 promoters by various mechanisms, L1s can capture novel cellular promoters by TPRT followed by incomplete reverse transcription (Fig. 9). This would then put the element under control of a new promoter. The divergent structures of primate and rodent elements support the idea that at least one such event occurred between rodent and primate lineages. The hypothesis that the promoters are independently derived yet retain germ-line- and neuron-specific activities could be tested by

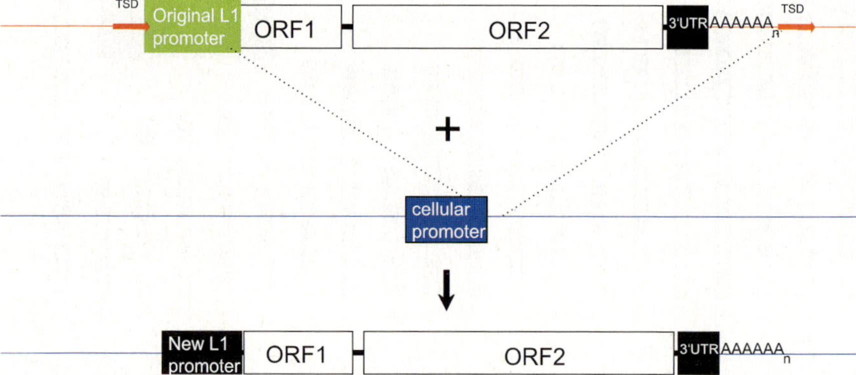

Fig. 9. Promoter capture model. L1 may capture cellular promoters during evolution by transposing a partially truncated element. During the TPRT reaction, the reverse transcription of L1 RNA may extend through ORF1 but fail to copy its own promoter. If this incomplete element is inserted downstream of a cellular promoter, then the L1 might capture this sequence as its own novel promoter

examining the retrotranspositional activity of native mouse elements or *ORFeus* driven by the native mouse L1 promoter. Such experiments are in progress (collaboration with A. Muotri and F. Gage).

References

An W, Han JS, Wheelan SJ, Davis ES, Coombes CE, Ye P, Triplett C, Boeke JD (2006) Active retrotransposition by a synthetic L1 element in mice. Proc Natl Acad Sci USA 103:18662–18667

Babushok DV, Ostertag EM, Courtney CE, Choi JM, Kazazian HH (2006) L1 integration in a transgenic mouse model. Genome Res 16:240–250

Branciforte D, Martin SL (1994) Developmental and cell type specificity of LINE-1 expression in mouse testis: implications for transposition. Mol Cell Biol 14:2584–2592

Chalker DL, Sandmeyer SB (1990) Transfer RNA genes are genomic targets for de novo transposition of the yeast retrotransposon Ty3. Genetics 126:837–850

Collier LS, Carlson CM, Ravimohan S, Dupuy AJ, Largaespada DA (2005) Cancer gene discovery in solid tumours using transposon-based somatic mutagenesis in the mouse. Nature 436:272–276

Cost GJ, Boeke JD (1998) Targeting of human retrotransposon integration is directed by the specificity of the L1 endonuclease for regions of unusual DNA structure. Biochemistry 37:18081–18093

Curcio MJ, Derbyshire KM (2003) The outs and ins of transposition: from mu to kangaroo. Nature Rev Mol Cell Biol 4:865–877

Dawkins R (1976) The selfish gene. Oxford University Press, Oxford

Devine SE, Boeke JD (1996) Integration of the yeast retrotransposon Ty1 is targeted to regions upstream of genes transcribed by RNA polymerase III. Genes Dev 10:620–633

Dewannieux M, Esnault C, Heidmann T (2003) LINE-mediated retrotransposition of marked Alu sequences. Nature Genet 35:41–48

Dupuy AJ, Akagi K, Largaespada DA, Copeland NG, Jenkins NA (2005) Mammalian mutagenesis using a highly mobile somatic Sleeping Beauty transposon system. Nature 436:221–226

Ergun S, Buschmann C, Heukeshoven J, Dammann K, Schnieders F, Lauke H, Chalajour F, Kilic N, Stratling WH, Schumann GG (2004) Cell type-specific expression of LINE-1 open reading frames 1 and 2 in fetal and adult human tissues. J Biol Chem 279:27753–27763

Esnault C, Maestre J, Heidmann T (2000) Human LINE retrotransposons generate processed pseudogenes. Nature Genet 24:363–367

Feng Q, Moran JV, Kazazian HH, Jr., Boeke JD (1996) Human L1 retrotransposon encodes a conserved endonuclease required for retrotransposition. Cell 87:905–916

Gibbs RA, Weinstock GM, Metzker ML, Muzny DM, Sodergren EJ, Scherer S, Scott G, Steffen D, Worley KC, Burch PE, Okwuonu G, Hines S, Lewis L, DeRamo C, Delgado O, Dugan-Rocha S, Miner G, Morgan M, Hawes A, Gill R, Celera, Holt RA, Adams MD, Amanatides PG, Baden-Tillson H, Barnstead M, Chin S, Evans CA, Ferriera S, Fosler C, Glodek A, Gu Z, Jennings D, Kraft CL, Nguyen T, Pfannkoch CM, Sitter C, Sutton GG, Venter JC, Woodage T, Smith D, Lee HM, Gustafson E, Cahill P, Kana A, Doucette-Stamm L, Weinstock K, Fechtel K, Weiss RB, Dunn DM, Green ED, Blakesley RW, Bouffard GG, De Jong PJ, Osoegawa K, Zhu B, Marra M, Schein J, Bosdet I, Fjell C, Jones S, Krzywinski M, Mathewson C, Siddiqui A, Wye N, McPherson J, Zhao S, Fraser CM, Shetty J, Shatsman S, Geer K, Chen Y, Abramzon S, Nierman WC, Havlak PH, Chen R, Durbin KJ, Egan A, Ren Y, Song XZ, Li B, Liu Y, Qin X, Cawley S, Cooney AJ, D'Souza LM, Martin K, Wu JQ, Gonzalez-Garay ML, Jackson AR, Kalafus KJ, McLeod MP, Milosavljevic A, Virk D, Volkov A, Wheeler DA, Zhang Z, Bailey JA, Eichler EE, Tuzun E, Birney E, Mongin E, Ureta-Vidal A, Woodwark C, Zdobnov E, Bork P, Suyama M, Torrents D, Alexandersson M, Trask BJ, Young JM, Huang H, Wang H, Xing H, Daniels S, Gietzen D, Schmidt J, Stevens K, Vitt U, Wingrove J, Camara F, Mar Alba M, Abril JF, Guigo R, Smit A, Dubchak I, Rubin EM, Couronne O, Poliakov A, Hubner N, Ganten D, Goesele C,

Hummel O, Kreitler T, Lee YA, Monti J, Schulz H, Zimdahl H, Himmelbauer H, Lehrach H, Jacob HJ, Bromberg S, Gullings-Handley J, Jensen-Seaman MI, Kwitek AE, Lazar J, Pasko D, Tonellato PJ, Twigger S, Ponting CP, Duarte JM, Rice S, Goodstadt L, Beatson SA, Emes RD, Winter EE, Webber C, Brandt P, Nyakatura G, Adetobi M, Chiaromonte F, Elnitski L, Eswara P, Hardison RC, Hou M, Kolbe D, Makova K, Miller W, Nekrutenko A, Riemer C, Schwartz S, Taylor J, Yang S, Zhang Y, Lindpaintner K, Andrews TD, Caccamo M, Clamp M, Clarke L, Curwen V, Durbin R, Eyras E, Searle SM, Cooper GM, Batzoglou S, Brudno M, Sidow A, Stone EA, Payseur BA, Bourque G, Lopez-Otin C, Puente XS, Chakrabarti K, Chatterji S, Dewey C, Pachter L, Bray N, Yap VB, Caspi A, Tesler G, Pevzner PA, Haussler D, Roskin KM, Baertsch R, Clawson H, Furey TS, Hinrichs AS, Karolchik D, Kent WJ, Rosenbloom KR, Trumbower H, Weirauch M, Cooper DN, Stenson PD, Ma B, Brent M, Arumugam M, Shteynberg D, Copley RR, Taylor MS, Riethman H, Mudunuri U, Peterson J, Guyer M, Felsenfeld A, Old S, Mockrin S.Collins F (2004) Genome sequence of the Brown Norway rat yields insights into mammalian evolution. Nature 428:493–521

Goodier JL, Ostertag EM, Du K, Kazazian HH, Jr. (2001) A novel active L1 retrotransposon subfamily in the mouse. Genome Res 11:1677–1685

Han JS, Boeke JD (2004) A highly active synthetic mammalian retrotransposon. Nature 429:314–318

Ji H, Moore DP, Blomberg MA, Braiterman LT, Voytas DF, Natsoulis G, Boeke JD (1993) Hotspots for unselected Ty1 transposition events on yeast chromosome III are near tRNA genes and LTR sequences. Cell 73:1007–1018

Khan H, Smit A, Boissinot S (2006) Molecular evolution and tempo of amplification of human LINE-1 retrotransposons since the origin of primates. Genome Res 16:78–87

Kim JM, Vanguri S, Boeke JD, Gabriel A, Voytas DF (1998) Transposable elements and genome organization: a comprehensive survey of retrotransposons revealed by the complete Saccharomyces cerevisiae genome sequence. Genome Res 8:464–478

Kirkness EF, Bafna V, Halpern AL, Levy S, Remington K, Rusch DB, Delcher AL, Pop M, Wang W, Fraser CM, Venter JC (2003) The dog genome: survey sequencing and comparative analysis. Science 301:1898–1903

Lander ES, Linton LM, Birren B, Nusbaum C, Zody MC, Baldwin J, Devon K, Dewar K, Doyle M, FitzHugh W, Funke R, Gage D, Harris K, Heaford A, Howland J, Kann L, Lehoczky J, LeVine R, McEwan P, McKernan K, Meldrim J, Mesirov JP, Miranda C, Morris W, Naylor J, Raymond C, Rosetti M, Santos R, Sheridan A, Sougnez C, Stange-Thomann N, Stojanovic N, Subramanian A, Wyman D, Rogers J, Sulston J, Ainscough R, Beck S, Bentley D, Burton J, Clee C, Carter N, Coulson A, Deadman R, Deloukas P, Dunham A, Dunham I, Durbin R, French L, Grafham D, Gregory S, Hubbard T, Humphray S, Hunt A, Jones M, Lloyd C, McMurray A, Matthews L, Mercer S, Milne S, Mullikin JC, Mungall A, Plumb R, Ross M, Shownkeen R, Sims S, Waterston RH, Wilson RK, Hillier LW, McPherson JD, Marra MA, Mardis ER, Fulton LA, Chinwalla AT, Pepin KH, Gish WR, Chissoe SL, Wendl MC, Delehaunty KD, Miner TL, Delehaunty A, Kramer JB, Cook LL, Fulton RS, Johnson DL, Minx PJ, Clifton SW, Hawkins T, Branscomb E, Predki P, Richardson P, Wenning S, Slezak T, Doggett N, Cheng JF, Olsen A, Lucas S, Elkin C, Uberbacher E, Frazier M, Gibbs RA, Muzny DM, Scherer SE, Bouck JB, Sodergren EJ, Worley KC, Rives CM, Gorrell JH, Metzker ML, Naylor SL, Kucherlapati RS, Nelson DL, Weinstock GM, Sakaki Y, Fujiyama A, Hattori M, Yada T, Toyoda A, Itoh T, Kawagoe C, Watanabe H, Totoki Y, Taylor T, Weissenbach J, Heilig R, Saurin W, Artiguenave F, Brottier P, Bruls T, Pelletier E, Robert C, Wincker P, Smith DR, Doucette-Stamm L, Rubenfield M, Weinstock K, Lee HM, Dubois J, Rosenthal A, Platzer M, Nyakatura G, Taudien S, Rump A, Yang H, Yu J, Wang J, Huang G, Gu J, Hood L, Rowen L, Madan A, Qin S, Davis RW, Federspiel NA, Abola AP, Proctor MJ, Myers RM, Schmutz J, Dickson M, Grimwood J, Cox DR, Olson MV, Kaul R, Shimizu N, Kawasaki K, Minoshima S, Evans GA, Athanasiou M, Schultz R, Roe BA, Chen F, Pan H, Ramser J, Lehrach H, Reinhardt R, McCombie WR, de la Bastide M, Dedhia N, Blocker H, Hornischer K, Nordsiek G, Agarwala R, Aravind L, Bailey JA, Bateman A, Batzoglou S, Birney E, Bork P, Brown DG, Burge CB, Cerutti L, Chen HC, Church D,

Clamp M, Copley RR, Doerks T, Eddy SR, Eichler EE, Furey TS, Galagan J, Gilbert JG, Harmon C, Hayashizaki Y, Haussler D, Hermjakob H, Hokamp K, Jang W, Johnson LS, Jones TA, Kasif S, Kaspryzk A, Kennedy S, Kent WJ, Kitts P, Koonin EV, Korf I, Kulp D, Lancet D, Lowe TM, McLysaght A, Mikkelsen T, Moran JV, Mulder N, Pollara VJ, Ponting CP, Schuler G, Schultz J, Slater G, Smit AF, Stupka E, Szustakowski J, Thierry-Mieg D, Thierry-Mieg J, Wagner L, Wallis J, Wheeler R, Williams A, Wolf YI, Wolfe KH, Yang SP, Yeh RF, Collins F, Guyer MS, Peterson J, Felsenfeld A, Wetterstrand KA, Patrinos A, Morgan MJ, Szustakowki J, de Jong P, Catanese JJ, Osoegawa K, Shizuya H, Choi S, Chen YJ (2001) Initial sequencing and analysis of the human genome. Nature 409:860–921
Luan DD, Korman MH, Jakubczak JL, Eickbush TH (1993) Reverse transcription of R2Bm RNA is primed by a nick at the chromosomal target site: a mechanism for non-LTR retrotransposition. Cell 72:595–605
Martin SL (1991) Ribonucleoprotein particles with LINE-1 RNA in mouse embryonal carcinoma cells. Mol Cell Biol 11:4804–4807
Mathias SL, Scott AF, Kazazian HH, Jr., Boeke JD, Gabriel A (1991) Reverse transcriptase encoded by a human transposable element. Science 254:1808–1810
Moran JV, Holmes SE, Naas TP, DeBerardinis RJ, Boeke JD, Kazazian HH, Jr. (1996) High frequency retrotransposition in cultured mammalian cells. Cell 87:917–927
Mulhardt C, Fischer M, Gass P, Simon-Chazottes D, Guenet JL, Kuhse J, Betz H, Becker CM (1994) The spastic mouse: aberrant splicing of glycine receptor beta subunit mRNA caused by intronic insertion of L1 element. Neuron 13:1003–1015
Muotri AR, Chu VT, Marchetto MC, Deng W, Moran JV, Gage FH (2005) Somatic mosaicism in neuronal precursor cells mediated by L1 retrotransposition. Nature 435:903–910
Padgett RW, Hutchison CA, 3rd, Edgell MH (1988) The F-type 5' motif of mouse L1 elements: a major class of L1 termini similar to the A-type in organization but unrelated in sequence. Nucleic Acids Res 16:739–749
Swergold GD (1990) Identification, characterization, and cell specificity of a human LINE-1 promoter. Mol Cell Biol 10:6718–6729.
Trelogan SA, Martin SL (1995) Tightly regulated, developmentally specific expression of the first open reading frame from LINE-1 during mouse embryogenesis. Proc Natl Acad Sci USA 92:1520–1524
Waterston RH, Lindblad-Toh K, Birney E, Rogers J, Abril JF, Agarwal P, Agarwala R, Ainscough R, Alexandersson M, An P, Antonarakis SE, Attwood J, Baertsch R, Bailey J, Barlow K, Beck S, Berry E, Birren B, Bloom T, Bork P, Botcherby M, Bray N, Brent MR, Brown DG, Brown SD, Bult C, Burton J, Butler J, Campbell RD, Carninci P, Cawley S, Chiaromonte F, Chinwalla AT, Church DM, Clamp M, Clee C, Collins FS, Cook LL, Copley RR, Coulson A, Couronne O, Cuff J, Curwen V, Cutts T, Daly M, David R, Davies J, Delehaunty KD, Deri J, Dermitzakis ET, Dewey C, Dickens NJ, Diekhans M, Dodge S, Dubchak I, Dunn DM, Eddy SR, Elnitski L, Emes RD, Eswara P, Eyras E, Felsenfeld A, Fewell GA, Flicek P, Foley K, Frankel WN, Fulton LA, Fulton RS, Furey TS, Gage D, Gibbs RA, Glusman G, Gnerre S, Goldman N, Goodstadt L, Grafham D, Graves TA, Green ED, Gregory S, Guigo R, Guyer M, Hardison RC, Haussler D, Hayashizaki Y, Hillier LW, Hinrichs A, Hlavina W, Holzer T, Hsu F, Hua A, Hubbard T, Hunt A, Jackson I, Jaffe DB, Johnson LS, Jones M, Jones TA, Joy A, Kamal M, Karlsson EK, Karolchik D, Kasprzyk A, Kawai J, Keibler E, Kells C, Kent WJ, Kirby A, Kolbe DL, Korf I, Kucherlapati RS, Kulbokas EJ, Kulp D, Landers T, Leger JP, Leonard S, Letunic I, Levine R, Li J, Li M, Lloyd C, Lucas S, Ma B, Maglott DR, Mardis ER, Matthews L, Mauceli E, Mayer JH, McCarthy M, McCombie WR, McLaren S, McLay K, McPherson JD, Meldrim J, Meredith B, Mesirov JP, Miller W, Miner TL, Mongin E, Montgomery KT, Morgan M, Mott R, Mullikin JC, Muzny DM, Nash WE, Nelson JO, Nhan MN, Nicol R, Ning Z, Nusbaum C, O'Connor MJ, Okazaki Y, Oliver K, Overton-Larty E, Pachter L, Parra G, Pepin KH, Peterson J, Pevzner P, Plumb R, Pohl CS, Poliakov A, Ponce TC, Ponting CP, Potter S, Quail M, Reymond A, Roe BA, Roskin KM, Rubin EM, Rust AG, Santos R, Sapojnikov V, Schultz B, Schultz J, Schwartz MS, Schwartz S, Scott C, Seaman S, Searle S, Sharpe T, Sheridan A, Shownkeen R, Sims S, Singer JB,

Slater G, Smit A, Smith DR, Spencer B, Stabenau A, Stange-Thomann N, Sugnet C, Suyama M, Tesler G, Thompson J, Torrents D, Trevaskis E, Tromp J, Ucla C, Ureta-Vidal A, Vinson JP, Von Niederhausern AC, Wade CM, Wall M, Weber RJ, Weiss RB, Wendl MC, West AP, Wetterstrand K, Wheeler R, Whelan S, Wierzbowski J, Willey D, Williams S, Wilson RK, Winter E, Worley KC, Wyman D, Yang S, Yang SP, Zdobnov EM, Zody MC, Lander ES (2002) Initial sequencing and comparative analysis of the mouse genome. Nature 420:520–562
Wheelan SJ, Scheifele LZ, Martinez-Murillo F, Irizarry RA, Boeke JD (2006) Transposon insertion site profiling chip (TIP-chip). Proc Natl Acad Sci USA 103:17632–17637
Zou S, Ke N, Kim JM, Voytas DF (1996) The Saccharomyces retrotransposon Ty5 integrates preferentially into regions of silent chromatin at the telomeres and mating loci. Genes Dev 10:634–645

Ancient Retrotransposons as Possible Remnants of the Primitive RNP World

Roland Ivanyi-Nagy[1] and *Jean-Luc Darlix*[1]

Summary

A recently discovered common trait of all eukaryotic organisms is a high prevalence of mobile genetic elements in their genome. From yeast to man, transposable elements are mostly represented by retroelements, which account for about half of the genome content in mammals and replicate by means of an obligatory RNA intermediate. According to the "RNA world" hypothesis, one step of the "prebiotic life" would correspond to populations of small RNA molecules forming increasingly complex RNA networks generating and piling up information. To reach this organizational level, RNA molecules probably needed to interact with small basic peptides, thus forming the backbone of ribonucleoparticles (RNPs). Aggregation and condensation of diverse RNA populations by basic peptides can generate arrays of large RNPs in which a drastic increase in RNA concentration leads to a crowding phenomenon facilitating RNA-RNA, RNA-peptide and peptide-peptide interactions and biochemical reactions. In addition, small basic peptides can have critical properties, known as RNA chaperoning, whereby the most stable structure of RNA molecules is rapidly reached and RNA-RNA annealing reactions can take place in a highly dynamic fashion. Thus, RNA chaperoning provided by basic peptides within highly crowded RNA populations may actively create novel information essential for the basic mechanisms of life. Simple retroelements from yeast to man encode such simple basic peptides with the RNA binding and chaperoning properties that are needed for their replication. Here we will briefly review RNA chaperones of retroelements and their mode of action, whereby these intrinsically flexible proteins can interact with many different partners to establish networks required for the replication of retroelements. We will also consider some mechanisms thought to regulate amplification of retroelements to control their invasion of and damage to eukaryotic genomes.

Simple Retrotransposons Compose Half of our Genome

Intensive genome sequencing has revealed that the retrotransposon class of retroelements can account for nearly half of the genome content in mammals (International

---

[1] LaboRetro, Unité de Virologie humaine INSERM, IFR128, ENS Lyon 46 allée d'Italie, 69364 Lyon, France
e-mail:jldarlix@ens-lyon.fr

Gage et al.
Retrotransposition, Diversity and the Brain
© Springer-Verlag Berlin Heidelberg 2008

Human Genome Sequencing Consortium 2001; Mouse Genome Sequencing Consortium 2002). Retroelements, and notably retrotransposons, are viewed as very ancient genetic structures because they are ubiquitous in eukaryotes, from yeast to human (Kazazian 2004). Replicating retroelements are small mobile genetic entities formed of protein coding sequences, or open reading frames (ORF), flanked by terminal repeats. Retroelements have been classified into retrotransposons with long terminal repeats (LTR retrotransposons) such as TYs of yeast, or retrotransposons without LTR (non-LTR retrotransposons) such as LINEs (Ostertag and Kazazian 2001; Kazazian 2004), and retroviruses that encompass the widespread simple gamma-retroviruses, notably the exogenous and endogenous murine leukemia viruses (MLV) and the family of complex lentiviruses best represented by the human immunodeficiency viruses (HIV) and simian immunodeficiency viruses (SIV)(Goff 2007; Balvay et al. 2007).

Retrotransposons and retroviruses share a similar genetic organization, firstly ORF1 or *gag* encoding the major structural core protein(s), and secondly ORF2 or *pol* coding for the enzymes that are the RNA/DNA-dependent DNA polymerase known as reverse transcriptase (RT) with RNaseH activity, and the integrase or endonuclease (IN/EN; Ostertag and Kazazian 2001; Kazazian 2004). In addition, LTR-retrotransposons and retroviruses code for a protease (PR) in *pol*. Replication-competent retroviruses also contain a third ORF, the envelope (*env*), which is necessary for the production of infectious virions and their dissemination in a horizontal fashion within animal populations and eventually between different populations (Goff 2007; Balvay et al. 2007).

The RNA/RNP World and Retroelements

RNA molecules are central players in all forms of life, from viroids to cells. Major classes of RNA molecules present in cells include transfer RNAs (tRNAs), small nuclear, nucleolar and cytoplasmic RNAs (sn-, sno- and scRNAs), ribosomal RNAs (rRNAs), pre- and messenger RNA (mRNAs) coding for proteins, and lastly, the recently discovered very large family of microRNAs, which includes small temporal RNAs (stRNAs), small interfering RNAs (siRNAs), and Piwi-interacting RNAs (piRNAs) that regulate the expression and stability of mRNAs (Lee and Ambros 2001; Lagos-Quintana et al. 2001; O'Driscoll 2006; Lau et al. 2006). These diverse RNAs act in a coordinated, highly regulated fashion as parts of complex RNA-RNA and RNA-protein interaction networks, regulating the flow of information from DNA to proteins. In addition, ribozymes – RNA molecules with catalytic activity – are considered to be vestiges of the hypothetical prebiotic RNA world (Gilbert 1986), when both information storage and chemical catalysis were carried out by RNA. As already pointed out, retroviral and retrotransposon RNAs are diverse and abundant, with a major impact on eukaryotic genome formation and dynamics (Kazazian 2004). Viroid RNAs are unusual because they are circular and single-stranded and contain a large number of modified nucleotides (Symons 1997).

In most instances RNAs are single-stranded and thus correspond to flexible macromolecules that can adopt a wide variety of alternative conformations. However, in many cases only one, well-defined structure is thought to be biologically active, which, besides folding kinetics and thermodynamics, is also influenced by interacting partner(s). At the same time, a large fraction of an RNA population can be trapped in incorrect structures, highlighting the "RNA folding problem" in a dynamic biological environment (Russell et al. 2002). Therefore, RNA folding must be assisted so that functionally relevant conformations can be rapidly reached. In addition, assistance must be provided so that possible RNA-RNA interactions can rapidly take place, leading to the formation of a network of interacting, functionally active molecules.

Replication of retroelements takes place in a very small, compact ribonucleoparticle (RNP) structure, where the single-stranded RNA is converted into a double-stranded DNA by RT, assisted by a basic RNA-binding peptide encoded by the retroelement. This unique basic peptide is encoded by ORF1 (Gag in retroviruses and certain LTR-retrotransposons), and is called nucleocapsid protein (NC; Darlix et al. 2007). In addition to their nucleic acid binding ability, retroviral NC and retrotransposon NC-like peptides have in common potent RNA condensing and chaperoning properties (Cristofari and Darlix 2002).

According to the RNA world scenario, RNA replicators constituted a prebiotic form of "life," serving at the same time as genetic material and catalysts, predating the appearance of templated protein synthesis (Gilbert 1986). Indeed, ribozymes in extant organisms and artificial catalytic RNA molecules testify to the catalytic versatility and capacity of RNA (e.g., Hager et al. 1996; Johnston et al. 2001; Spirin 2002; Muller 2006; Chen et al. 2007). However, the length of self-replicating RNA molecules and thus the complexity of the RNA world were greatly limited by the inherent low fidelity of (protein-unassisted) RNA replication, resulting in a lethal mutational load over a certain "genome" length (error threshold; Eigen 1971). Small RNA chaperone-like peptides, synthesized under prebiotic conditions in an untemplated manner, may have relaxed these limitations by stabilizing and folding RNA molecules and facilitating RNA-RNA interactions (Poole et al. 1998). In agreement with the proposed primeval origin of RNA chaperones, extant proteins with RNA chaperone activity frequently contain low amino acid complexity regions and they may function in the absence of a well-defined structure (see below; Tompa and Csermely 2004; Ivanyi-Nagy et al. 2005). In addition, even short unstructured peptides may have potent RNA chaperone activities (Ivanyi-Nagy and Darlix, unpublished data).

A second major increase in information content and complexity at the dawn of life could have occurred in an RNA world by the invention of reverse transcription, followed by the takeover of information storage by the more stable DNA. Given that retroelements are ancient, highly repetitive and dispersed genetic entities in all eukaryotic genomes, and that they replicate by reverse transcription of RNA into DNA, they have been suggested to be early players in the formation of complex DNA genomes (Forterre 2005; Koonin et al. 2006). As discussed below, RNA chaperones probably played an important role in this process, too.

A New World of Disorder

RNA chaperone proteins do not share a characteristic domain organization, specific motif or signature sequence that could provide insights into the mechanism of RNA structural rearrangements directed by them. Indeed, while some retroelement contain one or two copies of a specific RNA-binding motif, called CCHC-type zinc finger

($Cx_2Cx_4Hx_4C$), others – like TY1 or Gypsy – lack a well-defined, consensus RNA-binding motif (Fig. 1; Covey 1986; Cristofari et al. 2002; Gabus et al. 2006). Despite the absence of considerable sequence homology, NC or NC-like peptides are all rich in basic amino acids and proline, and they frequently contain low amino acid complexity regions, a combination of features that results in a highly flexible conformation (Dunker et al. 2001; Fig. 1). Such proteins and peptides are now referred to as intrinsically unstructured proteins (IUPs).

Based on the abundance of relatively long, continuous disordered fragments in RNA chaperones, Tompa and Csermely (2004) recently proposed an elegant model to account for the ATP-independent action of these proteins. According to their "entropy exchange model," cyclic RNA binding and release of the RNA chaperone, coupled with a reciprocal entropy transfer process between the RNA and the protein, would provide the driving force for rapid RNA structural rearrangements (Fig. 2). Based on that model, RNA chaperones contact RNA in a (partly) unstructured state (Fig. 2, step 1). RNA binding induces local disorder-to-order transition (folding) of the protein, accompanied by the partial melting of the RNA secondary structure (step 2). The energy required for RNA structure destabilization is thus covered by an entropy transfer associated with

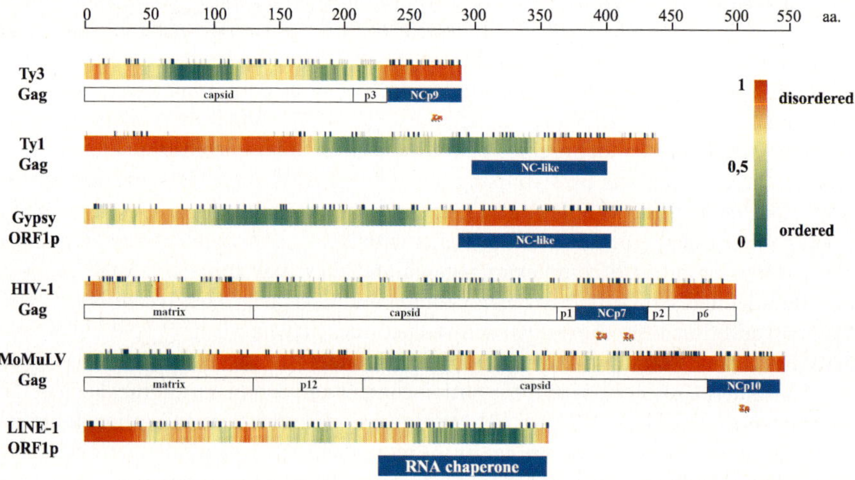

Fig. 1. Domain organization and intrinsic disorder in retroelement Gag proteins. Disordered regions in retrovirus and retrotransposon Gag proteins were predicted using the DisProt VL3-H predictor (http://www.ist.temple.edu/disprot/predictor.php). GenBank accession numbers are the following: Ty3 Gag Q12173; Ty1 Gag AAA66937; Gypsy ORF1p AAA70218; HIV-1 Gag NP_057850; MoMuLV Gag AAC82566; and LINE-1 ORF1p AAA67726. An amino acid with a disorder score above 0.5 is considered to be in a disordered environment; one below 0.5 is considered to be ordered. The predicted disorder is illustrated by a color scale, with highly flexible segments in *red* and well-folded domains in *green*. Basic and acidic amino acid residues are indicated by *dark blue* and *mauve symbols*, respectively. Zn stands for the CCHC-type zinc finger motif that is required for proper virion formation and genomic RNA replication in simple (i.e., MLVs) and complex retroviruses (i.e., HIV-1; reviewed in Darlix et al. 2007). Note that the NC peptidic regions (in *dark blue*) are small and flexible, in particular those of the ancient retrotransposons TY1, TY3 and Gypsy

RNA binding. Finally, the RNA chaperone is released from the partially melted RNA, which is free to resume a conformational search (step 3). Successive cycles of transient binding, accompanied by disorder-to-order transitions, lead to the formation of the most stable RNA secondary structure (Fig. 2).

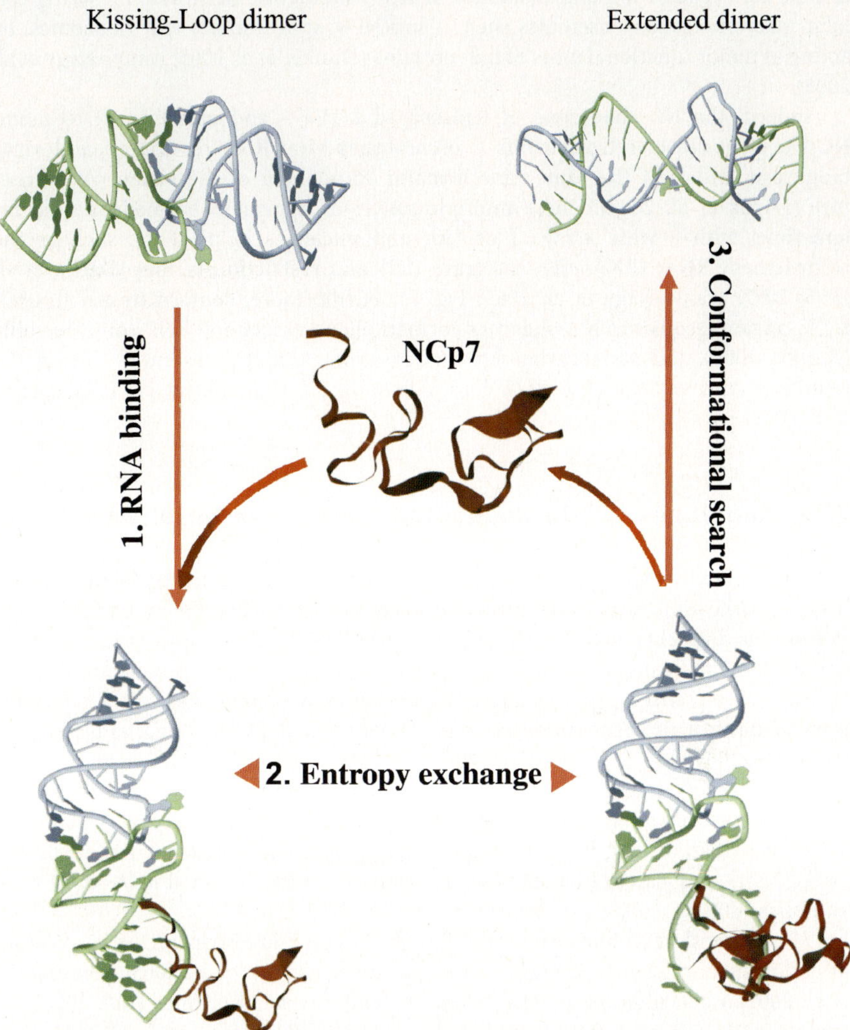

Fig. 2. The entropy exchange model of RNA chaperone function. The entropy exchange model of Tompa and Csermely (2004) is illustrated here by the example of NCp7-mediated HIV-1 genomic RNA dimerization. In the absence of NCp7, the dimer initiation sequence (DIS) of HIV-1 forms a kissing-loop dimer structure. Upon binding of the highly flexible protein (*step 1*), the RNA structure is destabilized (*step 2*), and adopts the more stable, extended dimer conformation (*step 3*). See more detailed explanation of the model in the text. The figure is not drawn to scale. PDB entries 1JU1, 2F4X, and 1ESK were used for figure drawing

In addition to providing a mechanistic basis for chaperone function, intrinsic disorder may confer numerous additional advantages for RNA chaperones of retroelements. Flexibility of the peptide chain allows high-affinity interactions with multiple, structurally diverse substrate molecules (one-to-many signaling; Kriwacki et al. 1996; Dunker et al. 1998). In this way, disordered peptidic segments may play a central role in the organization and dynamics of RNA interaction networks, including cellular macromolecular ensembles such as hnRNP's, spliceosomes and ribosomes, by acting as major functional units of hub proteins (Dunker et al. 2005; Ivanyi-Nagy et al. 2005).

Indeed, the NC chaperone of widespread MLVs – and probably all NCs and NC-like peptides of retroelements – orchestrates viral RNA replication and virus biogenesis and is at the same time a major component of the virion core structure (Darlix et al. 2007). These multiple functions appear to be mediated via interactions with a wide array of protein and nucleic acid partners, such as the retroelement RNA, tRNAs, RT, integrase (IN) and restriction factors (Darlix et al. 1995, 2007; Ivanyi-Nagy et al. 2005; Fig. 3). Furthermore, conformational flexibility is advantageous for NC-oligomer formation, necessary for RNP-core assembly (Namba 2001) and stabilization by condensation, and it is essential during the multistep conversion of the RNA into cDNA by RT (Morellet et al. 1994; Darlix et al. 2007).

RNA Chaperones and the Regulation of Retrotransposon Replication

Given that retroelements can replicate at fairly high levels, their unregulated and continuous replication would create genetic instability in eukaryotes by damaging the genome via different mechanisms, such as insertion mutagenesis, gene inactivation, or chromosome rearrangement by homologous recombination between copies of retrotransposon sequences (Ostertag and Kazazian 2001). A genome-wide impact of retrotransposons has also been proposed to affect gene transcription in higher eukaryotes (Han et al. 2004).

However, retrotransposition appears to be tightly regulated in a cumulative mode, first at the transcriptional level, whereby transcription is either silenced by DNA methylation or halted, at least in part, by *cis*-acting retrotransposon sequences (Han and Boeke 2004), and also by host-encoded restriction factors and small RNAs (O'Donnel and Boeke 2007).

More recently, cytidine deaminases have been found to restrict retrotransposition of endogenous retroviruses and retrotransposons (Esnault et al. 2005; Schumacher et al. 2005; Muckenfuss et al. 2006; Stenglein and Harris 2006; Chiu et al. 2006). By analogy with HIV-1 restriction mediated by APOBEC3G (Harris et al. 2003; Mariani et al. 2003), this family of enzymes seems to operate by deaminating C residues in the newly made cDNA and thus through interaction with the reverse transcription RNA-NC-RT complex (Fig. 3). Another negative regulation provided by the APOBEC enzymes would be via specific RNPs interacting with the retroelement core (Chiu et al. 2006; Stenglein and Harris 2006; Gallois-Montbrun et al. 2007; Holmes et al. 2007).

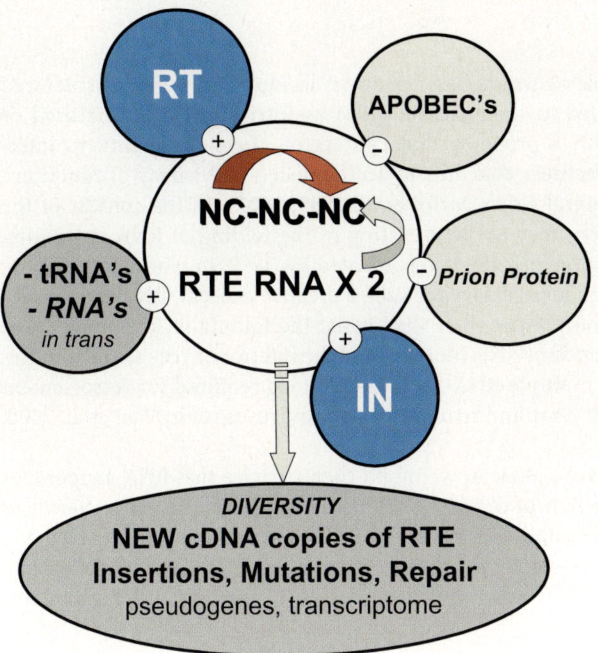

Fig. 3. Network of NC chaperone-directed interactions in retroelements. The simple scheme illustrates: (1) formation of the RNP or Core structure of the Retroelement (RTE) upon binding of NC molecules to the the homologous RTE RNA (see *arrows*); (2) core formation and condensation induce molecular crowding and provide protection to the structure; (3) recruitment of two RTE RNA molecules by NC (*arrows*), which chaperones RNA dimerization, as shown for TYs and retroviruses; (4) packaging of tRNAs that serve as primers for cDNA synthesis by RT chaperoned by NC; (5) interactions between RTE core structure and RNPs containing APOBECs can potentially change C > U in the RTE RNA and in the newly made cDNA, causing an efficient restriction of RTE replication and dissemination as well as microdiversity; (6) the prion protein might also interfere with the process of cDNA synthesis and more generally with the assembly of retroviral particles; (7) reverse transcription of the dimeric RNA – or pseudodiploid RNA – can generate RTE diversity by means of recombination reactions; (8) RNAs other than RTE can be packaged in *trans* and reverse transcribed, as shown for mouse sarcoma and leukemia viruses; (9) Integration of the newly made cDNA copy is mediated by the integrase (IN) or endonuclease enzyme, which can result in mutational insertions or promoter activation with an impact on the transcriptional pattern of the host genome (transcriptome)

Furthermore, the prion protein, PrP, may well be a restriction factor acting during core assembly. PrP resembles retroelement NC since it has potent RNA binding and chaperoning properties (Gabus et al. 2001a,b). Upon binding to the retroelement RNA, PrP chaperones its dimerization in vitro, resulting in the formation of highly compact RNPs (Gabus et al. 2001b). In the context of retrovirus-infected cells, PrP is recruited into newly formed viral particles and at the same time inhibits virion production and infectivity (Leblanc et al. 2004, 2006). It remains to be shown whether PrP also has a negative impact on retrotransposition.

Conclusions

RNA chaperones of retroelements are basic peptides lacking well-defined structured motifs. This finding has led to their classification as intrinsically unstructured or disordered proteins (IUPs), a property that confers on them the ability to interact with many different partners and thus to accomplish a wide array of functions, as demonstrated for retroviral NCs (Darlix et al. 1995, 2007). In the context of the disorder-to-order transition, they act by assisting in the folding of RNA molecules. Given the hypothesis that prebiotic life is represented by the RNA world and that the disorder-to-order transition is necessary for such a creative process, RNA chaperones of retroelements are central players since they assist the folding of RNA molecules, cause a crowding phenomenon of RNA molecules in the interior of very small volumes and provide assistance to multiple RNA-RNA interactions required for retroelement replication, such as TY1 of yeast and mammalian retroviruses (Cristofari et al. 2000, 2002; Darlix et al. 2007) .

Since retrotransposons are ancient, we might then surmise that RNA chaperones of retroelements are the origin of the more sophisticated, large cellular nucleic acid chaperones – such as p53, YB1 and FMRP - that are abundant proteins found in all living organisms, where they are essential in chromosome maintenance, DNA transcription, RNA splicing, trafficking, translation and degradation (Cristofari and Darlix 2002; Schroeder et al. 2004).

Acknowledgements. Work in LaboRetro is supported by ANRS, CNRS, INSERM, EC (TRIoH consortium) and Sidaction. RIN is supported by an ANRS fellowship.

References

Balvay L, Lopez Lastra M, Sargueil B, Darlix JL, Ohlmann T (2007) Translational control of retroviruses. Nature Rev Microbiol 5:128–140

Chen X, Li N, Ellington AD (2007) Ribozyme catalysis of metabolism in the RNA world. Chem Biodivers 4:633–655

Chiu YL, Witkowska HE, Hall SC, Santiago M, Soros VB, Esnault C, Heidmann T, Greene WC (2006) High-molecular-mass APOBEC3G complexes restrict Alu retrotransposition. Proc Natl Acad Sci USA 103:15588–15593

Covey SN (1986) Amino acid sequence homology in gag region of reverse transcribing elements and the coat protein gene of cauliflower mosaic virus. Nucleic Acids Res 14:623–633

Cristofari G, Darlix JL (2002) The ubiquitous nature of RNA chaperone proteins. Prog Nucleic Acid Res Mol Biol 72:223–268

Cristofari G, Ficheux D, Darlix JL (2000) The Gag-like protein of the yeast Ty1 retrotransposon contains a nucleic acid chaperone domain analogous to retroviral nucleocapsid proteins. J Biol Chem 275:19210–19217

Cristofari G, Bampi C, Wilhelm M, Wilhelm FX, Darlix JL (2002) A 5'-3' long-range interaction in Ty1 RNA controls its reverse transcription and retrotransposition. EMBO J 21:4368–4379

Darlix JL, Lapadat-Tapolsky M, de Rocquigny H, Roques B (1995) First glimpses at structure-function relationships of the nucleocapsid protein of retroviruses. J Mol Biol 254:523–537

Darlix JL, Garrido JL, Morellet N, Mély Y, de Rocquigny H (2007) Properties, functions and drug targeting of the multifunctional NC protein of HIV. Adv Pharmacol 55:299–346

Dunker AK, Garner E, Guilliot S, Romero P, Albrecht K, Hart J, Obradovic Z, Kissinger C, Villafranca JE (1998) Protein disorder and the evolution of molecular recognition: theory, predictions and observations. Pac Symp Biocomput 473–484

Dunker AK, Lawson JD, Brown CJ, Williams RM, Romero P, Oh JS, Oldfield CJ, Campen AM, Ratliff CM, Hipps KW, Ausio J, Nissen MS, Reeves R, Kang C, Kissinger CR, Bailey RW, Griswold MD, Chiu W, Garner EC, Obradovic Z (2001) Intrinsically disordered protein. J Mol Graph Model 19:26–59

Dunker AK, Cortese MS, Romero P, Iakoucheva LM, Uversky VN (2005) Flexible nets. The roles of intrinsic disorder in protein interaction networks. FEBS J 272:5129–5148

Eigen M (1971) Self organization of matter and the evolution of biological macromolecules. Naturwissenschaften 58:465–523

Esnault C, Heidmann O, Delebecque F, Dewannieux M, Ribet D, Hance AJ, Heidmann T, Schwartz O (2005) APOBEC3G cytidine deaminase inhibits retrotransposition of endogenous retroviruses. Nature 433:430–433

Forterre P (2005) The two ages of the RNA world, and the transition to the DNA world: a story of viruses and cells. Biochimie 87:793–803

Gabus C, Auxilien S, Pechoux C, Dormont D, Swietnicki W, Morillas M, Surewicz W, Nandi P, Darlix JL (2001a) The prion protein has DNA strand transfer properties similar to retroviral nucleocapsid protein. J Mol Biol 307:1011–1021

Gabus C, Derrington E, Leblanc P, Chnaiderman J, Dormont D, Swietnicki W, Morillas M, Surewicz WK, Marc D, Nandi P, Darlix JL (2001b) The prion protein has RNA binding and chaperoning properties characteristic of nucleocapsid protein NCp7 of HIV-1. J Biol Chem 276:19301–19309

Gabus C, Ivanyi-Nagy R, Depollier J, Bucheton A, Pelisson A, Darlix JL (2006) Characterization of a nucleocapsid-like region and of two distinct primer tRNA$^{\text{Lys},2}$ binding sites in the endogenous retrovirus Gypsy. Nucleic Acids Res 34:5764–5777

Gallois-Montbrun S, Kramer B, Swanson CM, Byers H, Lynham S, Ward M, Malim MH (2007) Antiviral protein APOBEC3G localizes to ribonucleoprotein complexes found in P bodies and stress granules. J Virol 81:2165–2178

Gilbert W (1986) The RNA world. Nature 319:618

Goff SP (2007) Host factors exploited by retroviruses. Nature Rev Microbiol 5:253–263

Hager AJ, Pollard JD, Szostak JW (1996) Ribozymes: aiming at RNA replication and protein synthesis. Chem Biol 3:717–725

Han JS, Boeke JD (2004) A highly active synthetic mammalian retrotransposon. Nature 429:314–318

Han JS, Szak ST, Boeke JD (2004) Transcriptional disruption by the L1 retrotransposon and implications for mammalian transcriptomes. Nature 429:268–274

Harris RS, Bishop KN, Sheehy AM, Craig HM, Petersen-Mahrt SK, Watt IN, Neuberger MS, Malim MH (2003) DNA deamination mediates innate immunity to retroviral infection. Cell 113:803–809. Erratum in: Cell 116:629

Holmes RK, Malim MH, Bishop KN (2007) APOBEC-mediated viral restriction: not simply editing? Trends Biochem Sci 32:118–128

International Human Genome Sequencing Consortium (2001) Initial sequencing and analysis of the human genome. Nature 409:860–921

Ivanyi-Nagy R, Davidovic L, Khandjian EW, Darlix JL (2005) Disordered RNA chaperone proteins: from functions to disease. Cell Mol Life Sci 62:1409–1417

Johnston WK, Unrau PJ, Lawrence MS, Glasner ME, Bartel DP (2001) RNA-catalyzed RNA polymerization: accurate and general RNA-templated primer extension. Science 292:1319–1325

Kazazian HH (2004) Mobile elements: drivers of genome evolution. Science 303:1626–1632

Koonin EV, Senkevich TG, Dolja VV (2006) The ancient Virus World and evolution of cells. Biol Direct 1:29

Kriwacki RW, Hengst L, Tennant L, Reed SI, Wright PE (1996) Structural studies of p21[Waf1/Cip1/Sdi1] in the free and Cdk2-bound state: conformational disorder mediates binding diversity. Proc Natl Acad Sci USA 93:11504–11509

Lagos-Quintana M, Rauhut R, Lendeckal W, Tuschl T (2001) Identification of novel genes coding for small expressed RNAs. Science 294:853–858

Lau NC, Seto AG, Kim J, Kuramochi-Miyagawa S, Nakano T, Bartel DP, Kingston RE (2006) Characterization of the piRNA complex from rat testes. Science 313:363–367

Leblanc P, Baas D, Darlix JL (2004) Analysis of the interactions between HIV-1 and the cellular prion protein in a human cell line. J Mol Biol 337:1035–1051

Leblanc P, Alais S, Porto-Carreiro I, Lehmann S, Grassi J, Raposo G, Darlix JL (2006) Retrovirus infection strongly enhances scrapie infectivity release in cell culture. EMBO J 25:2674–2685

Lee RC, Ambros V (2001) An extensive class of small RNAs in Caenorhabditis elegans. Science 294:862–864

Mariani R, Chen D, Schrofelbauer B, Navarro F, Konig R, Bollman B, Munk C, Nymark-McMahon H, Landau NR (2003) Species-specific exclusion of APOBEC3G from HIV-1 virions by Vif. Cell 114:21–31

Morellet N, de Rocquigny H, Mely Y, Jullian N, Demene H, Ottmann M, Gerard D, Darlix JL, Fournie-Zaluski MC, Roques BP (1994) Conformational behaviour of the active and inactive forms of the nucleocapsid NCp7 of HIV-1 studied by 1H NMR. J Mol Biol 235:287–301

Mouse Genome Sequencing Consortium (2002) Initial sequencing and comparative analysis of the mouse genome. Nature 420:520–562

Muckenfuss H, Hamdorf M, Held U, Perkovic M, Lower J, Cichutek K, Flory E, Schumann GG, Munk C (2006) APOBEC3 proteins inhibit human LINE-1 retrotransposition. J Biol Chem 281:22161–22172

Muller UF (2006) Re-creating an RNA world. Cell Mol Life Sci 63:1278–1293

Namba K (2001) Roles of partly unfolded conformations in macromolecular self-assembly. Genes Cells 6:1–12

O'Donnell KA, Boeke JD (2007) Mighty Piwis defend the germline against genome intruders. Cell 129:37–44.

O'Driscoll L (2006) The emerging world of microRNAs. Anticancer Res 26:4271–4278

Ostertag EM, Kazazian HH (2001) Biology of mammalian L1 retrotransposons. Annu Rev Genet 35:501–538

Poole AM, Jeffares DC, Penny D (1998) The path from the RNA world. J Mol Evol 46:1–17

Russell R, Zhuang X, Babcock HP, Millett IS, Doniach S, Chu S, Herschlag D (2002) Exploring the folding landscape of a structured RNA. Proc Natl Acad Sci USA 99:155–160

Schroeder R, Barta A, Semrad K (2004) Strategies for RNA folding and assembly. Nature Rev Mol Cell Biol 5:908–919

Schumacher AJ, Nissley DV, Harris RS (2005) APOBEC3G hypermutates genomic DNA and inhibits Ty1 retrotransposition in yeast. Proc Natl Acad Sci USA 102:9854–9859

Spirin AS (2002) Omnipotent RNA. FEBS Lett 530:4–8

Stenglein MD, Harris RS (2006) APOBEC3B and APOBEC3F inhibit L1 retrotransposition by a DNA deamination-independent mechanism. J Biol Chem 281:16837–16841

Symons RH (1997) Plant pathogenic RNAs and RNA catalysis. Nucleic Acids Res 25:2683–2689

Tompa P, Csermely P (2004) The role of structural disorder in the function of RNA and protein chaperones. FASEB J 18:1169–1175

Human Diversity and L1 Retrotransposon Biology: Creation of New Genes and Individual Variation in Retrotransposition Potential

H.H. Kazazian, Jr.[1], *M.d.C. Seleme*[1], *D.V. Babushok*[1], *D.M. Ostertag*[1], and *M.R. Vetter*[1], and *P.K. Mandal*[1]

The L1 retrotransposon is a major component of mammalian genomes and has molded them throughout evolution in many ways, thereby expanding the possibilities for human diversity. In this paper, we discuss one further mechanism by which L1 can alter the genome, namely the retrotransposition of a transcript involving sequences from two adjacent genes to form a new gene. In addition, a small fraction of L1 elements in the human genome is still actively retrotransposing, but there are little data on the extent of variation in retrotransposition potential among different individual human beings. Here we present evidence for considerable individual variation in L1 retrotransposition capability, a finding that has significant implications for the role of retrotransposition in present-day human neurological diversity.

Introduction

Roughly 500 000 mostly truncated copies of L1 retrotransposons occupy the human genome, of which about 6 000 are 6 kb or full length (Lander et al. 2001; Fig. 1). These full-length L1s have a 5'UTR containing both sense and antisense internal promoters (Swergold 1990; Speek 2001), a 1 kb ORF1 that encodes an RNA-binding protein with nucleic acid chaperone activity (Martin and Bushman 2001), a 4 kb ORF2 that encodes endonuclease activity (Feng et al. 1996), reverse transcriptase activity (Mathias et al. 1991), and a zinc knuckle domain (Fanning and Singer 1987). ORF2 is followed by a short 3' UTR and a poly A tail. Even though the complete functions of the ORFs are still unknown, both are critical for the retrotransposition process (Moran et al. 1996). Upon insertion of an L1 by target-primed reverse transcription (TPRT; Luan et al. 1993) on genomic DNA, the element is surrounded by 6–20 bp target site duplications of genomic sequence derived from the insertion site (Singer et al. 1993; Scott et al. 1987).

Fig. 1. Structure of a full-length 6 kb human L1 retrotransposon. See text for details

[1] Department of Genetics, University of Pennsylvania School of Medicine, Philadelphia, PA 19104, USA, e-mail: kazazian@mail.med.upenn.edu

Gage et al.
Retrotransposition, Diversity and the Brain
© Springer-Verlag Berlin Heidelberg 2008

L1s have contributed to genome development in a number of ways (Kazazian 2004). They are able to not only insert themselves (Ostertag and Kazazian 2001) but they also drive the insertion of Alu elements (Batzer and Deininger 2002), SVA elements (Ostertag et al. 2003; Wang et al. 2005), and processed pseudogenes (Esnault et al. 2000; Wei et al. 2001). Insertions of these other elements account for another 12% of the human genome (Zhang et al. 2003). Among additional ways that L1s have altered the genome are the 3' transduction of sequences from one site to another (Moran et al. 1999; Goodier et al. 2000; Pickeral et al. 2000), endonuclease-independent insertion (Morrish et al. 2001), formation of new exons (Gotea and Makalowski 2006), homologous recombination between repeat sequences (Deininger and Batzer 1999; Jurka 2004), and chimeric insertions involving small RNAs (Buzdin et al., 2003).

In the average human genome, roughly 100 L1s are still active, but the bulk of this retrotransposition capability resides in a small subset of these elements (Brouha et al. 2003). Here we first discuss another way in which L1 contributes to genome development and then the extent of individual variability in retrotransposition capability.

A New Mechanism for Gene Creation

Retrotransposition of mRNA to form processed pseudogenes is well known and has accounted for over 8,000 mRNA copies in the human genome (Esnault et al. 2000; Zhang et al. 2003). About 5% of these processed pseudogenes are transcribed due to serendipitous use of an external promoter (Zheng et al. 2005). At least one retrotransposed mRNA copy has landed in an intron during primate evolution and been incorporated into a fusion gene (Sayah et al., 2004). We have recently found another type of retrotransposed mRNA. In this case, the RNA is derived from a readthrough transcript connecting the RNA of a lipid kinase gene (gene 1-*PIP5K1A*) and a ubiquitin-binding protein gene (gene 2-*S5a*) located 6 kb apart. RNA splicing occurred between exon 13 of gene 1 and exon 2 of gene 2 (Fig. 2; Akiva et al. 2005; Babushok et al. 2007). The chimeric transcript was then retrotransposed into a new chromosomal site, and we call this gene *S5aL* for *S5a*-like. This event occurred roughly 17 million years ago, since this retrotransposed sequence is present in the genome of human beings, chimps, gorillas, and orangutans, but not in the rhesus genome. The splicing event maintained the reading frame such that translation occurs through to the stop codon of gene 2.

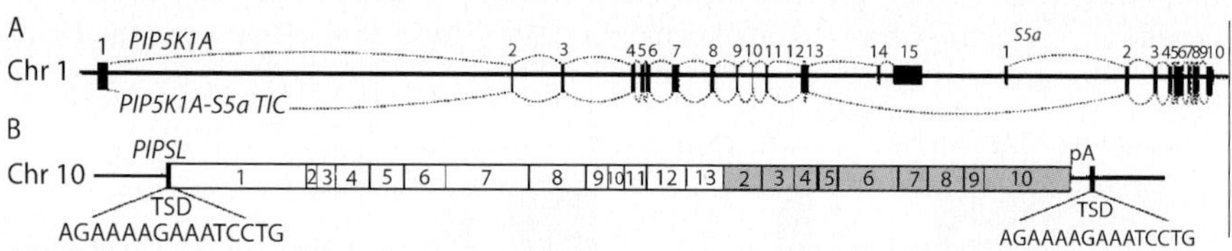

Fig. 2. Creation of a new gene by aberrant splicing and retrotransposition. (**A**) Neighboring 15-exon PIP5K1A and 10-exon S5a genes on chr 1 are spliced to form PIP5K1A, S5a, and PIP5K1A-S5a TIC mRNAs. *Black rectangles*, exons; *curved lines*, splicing. (**B**) PIP5K1A-S5a TIC was retrotransposed by L1 to create S5aL gene on chr 10. TSD, target site duplication; pA, A-rich repeat. Regions corresponding to PIP5K1A and S5a exons are shown as *white* and *gray rectangles*, respectively

S5aL is transcribed specifically in the testes in both humans and chimpanzees and is post-transcriptionally repressed by two independent mechanisms in these primate lineages. In humans, a nucleotide deletion within 50 nucleotides 3' of the initiation AUG codon of gene 1 knocks out translation. In the chimpanzee, the reading frame is intact, but translation is repressed by an unknown mechanism. Strong positive selection on S5aL has led to its rapid divergence from the parental genes PIP5K1A and S5a, forming a chimeric protein with a distinct cellular localization and minimal lipid kinase activity, but significant affinity for cellular ubiquitinated proteins.

Thus, S5aL is a tightly regulated, testes-specific, novel ubiquitin-binding protein that was formed by an unusual exon-shuffling mechanism in hominoid primates and may represent an example of rapid evolution of male reproductive genes that drives reproductive isolation and species divergence. To our knowledge, this is the only known example of an exon-shuffled, retrotransposed gene involving the mRNA of two genes in close proximity to each other. This fusion gene was found through a search of the human working draft sequence looking specifically for fusion genes; in a similar search of the mouse genome sequence, no such fusion genes were found. Whether any similar chimeric genes are present in other organisms is unknown at this time.

Extensive Individual Variation in Retrotransposon Capability Contributes to Human Genomic Diversity

Muotri et al. (2005) have hypothesized that human diversity in behavior and learning may be due in part to retrotransposition of L1 elements during neuronal development. They have found that human L1s can retrotranspose in neuronal precursors in cell culture and in cells contributing to various areas of the brain during mouse development. We had previously shown that the bulk of retrotransposition capability present in the human genome working draft is present in only a handful of L1 elements (Brouha et al. 2003). Since the extent to which human diversity may be controlled by retrotransposition depends heavily on the number and activity of active L1s in individual human genomes (in addition to unknown host factors), we decided to determine the extent of variation in retrotransposition capability for those six highly active or "hot" L1s among individual human beings. To carry out this analysis, we determined the number of copies of each element in each genome (homozygous present, heterozygous present, or absent), the nucleotide polymorphisms in each element, and the retrotransposition activity of each allele of the various elements (Seleme et al. 2006).

Of the six "hot" L1s that we had found in the database, we could not analyze three because two were extremely rare in human genomes and one was in highly repetitive DNA. Thus, our analysis centered on the remaining three "hot" L1s that we called L1A, L1B, and L1C. We first determined the presence/absence of polymorphism of the three L1s in 161 to 206 haploid genomes. Although the gene frequencies varied from population to population, none of the elements departed significantly from the Hardy-Weinberg equilibrium in any population. Overall, the frequencies of L1A, L1B and L1C were 0.19, 0.46, and 0.46, respectively. Since all three of these L1s belong to the youngest subfamily of L1s (Ta-1), it was not surprising that they are relatively recent additions to the human genome and still have fairly low gene frequencies.

We then obtained the complete sequence on L1A in 35 of 40 genomes containing this L1, on L1B in 59 of 67 genomes, and on L1C in 78 of 96 genomes. We found a previously uncharacterized allele in every five genomes for L1A and in every three genomes for L1B and L1C. L1A had 17 polymorphic sites among nine alleles, whereas L1B and L1C had 19 and 26 polymorphic sites among 18 and 26 alleles, respectively. Among the unexpected findings were four different nonsense mutations, two in a common allele of L1A and two in separate alleles of L1C.

Allelic Variation in Retrotransposition Capability: Hot and Cool Alleles

We tested the retrotransposition capability of 46 of the 52 alleles of L1A, L1B, and L1C in our cell culture assay. For the three elements, 22 of 46 tested alleles had 25% or greater retrotransposition activity compared with the reference L1 ($L1_{RP}$) and were called hot. However, the remaining 24 alleles (5 of 8 L1A, 3 of 16 L1B, and 16 of 22 L1C) had activity <25% of $L1_{RP}$ and were cool. Of all loci containing hot L1s, 33% (57 of 170) were cool.

L1 Retrotransposition Potential in Individuals and Populations

The number of possible genotypes in an individual for a locus with n alleles is $n(n+1)/2$ or 36 for L1A, 171 for L1B, and 351 for L1C with 8, 18, and 26 alleles, respectively. Because we designated certain activity categories and there were only three or four activity categories identified per locus, the many possible genotypes reduced to only 6, 10, and 9 possible activity phenotypes, respectively. Of those possible phenotypes, 66%, 80%, and 44% per locus corresponded to hot L1 phenotypes, defined as having a bi-allelic activity >25% that of $L1_{RP}$. The remaining phenotypes with bi-allelic activity <25% of $L1_{RP}$ were defined as cool L1 phenotypes. After assigning the activity values to the allelic variants that each individual carried at L1A, L1B, and L1C loci, we observed that only 11% (9 of 80), 44% (35 of 79) and 45% (36 of 80) of phenotypes per locus were hot.

For each individual, we added the activity values per locus to obtain the total L1 activity potential (three L1s combined). Figure 3 shows the wide distribution in L1 activity potentials per individual (from 0% to >300%) in the four populations that we sampled. Of the 80 individuals studied, one was excluded because we could not isolate his L1B element. For the remaining 79 individuals, 18% did not have a total L1 phenotype that was hot, 56% had a hot phenotype between 25% and 200%, and 26% had a very hot phenotype that was >200%. Thus, the data suggest that nearly half of individuals fall at the extremes of distribution of retrotransposition capability of these elements, suggesting that individuals vary significantly in their risk of a new insertion during meiosis or during development of their offspring.

To obtain an overall L1 activity potential per population, we added the values from all individuals in a population and divided by the number of individuals (Fig. 4). We tested whether the different populations were statistically different in their overall L1 activity potential. There was a >2-fold difference between the relative activity potential

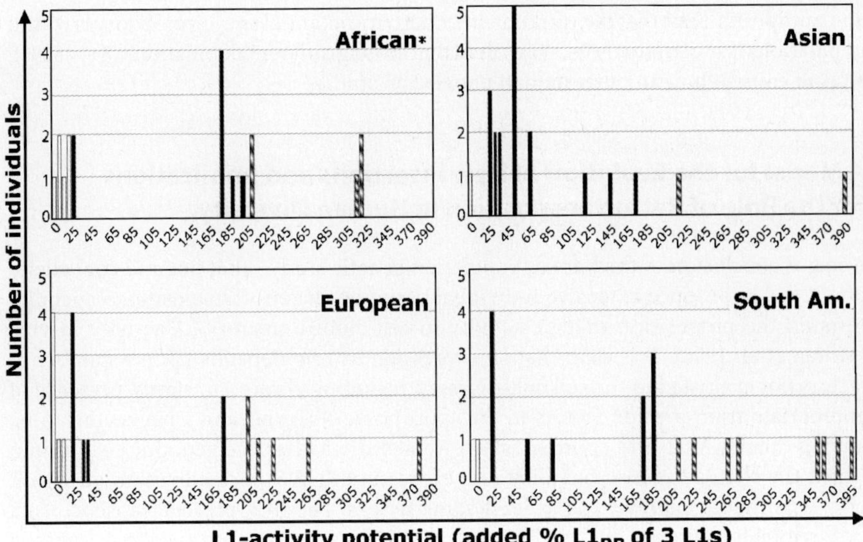

Fig. 3. Combined retrotransposition potential of three hot L1s/individual in four populations. From 26% (African) to 55% (South American) of individuals per population have a unique L1 activity potential. *White*, *black*, and *hatched* bars represent individuals lacking a hot L1 phenotype (<25%), having an intermediate L1 activity, and having a high L1 activity (>200%), respectively. * The African distribution is based on 19 individuals

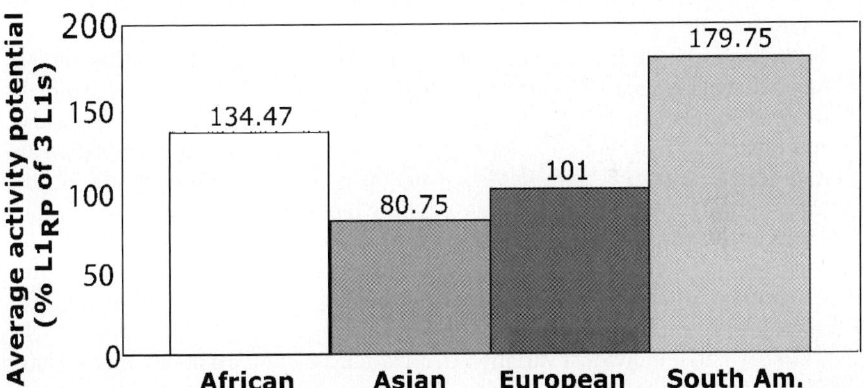

Fig. 4. Average retrotransposition potential of 3 hot L1s in four populations. The total retrotransposition potential of L1A, L1B, and L1C for each individual was divided by the number of individuals in the population to determine the average retrotransposition potential in each population. The means of the four populations are significantly different by an ANOVA test (p = 0.036)

of the highest (South Americans, 180%) and the lowest group (Asians, 81%). The hypothesis that all population means are equal was marginally rejected by an ANOVA test (P = 0.036) with South American and African means differing from those of Asians and Europeans. Note that the variation in L1 activity potential among individuals within populations is much larger (0–300%) than that among different populations (81–180%), a result consistent with other human population studies.

A Model for the Evolution of New Insertions and Implications for the Role of Retrotransposition in Human Diversity

A major conclusion of the human population genetic study - that hot and cool alleles of active L1s produce extensive individual variation in retrotransposition capability rests on the proposition that L1 activity in cell culture mirrors L1 activity in vivo (Moran et al. 1996). It is clear that L1 expression in vivo depends upon a number of factors not evaluated in the cell culture assay, including chromatin status, presence of appropriate transcription factors in the appropriate cell type, and DNA methylation, among others. As a first approximation we asked whether the genomic region into which the element is inserted allows its expression. L1A is located within an intron of a gene that is expressed in at least some tissues. For L1B, presence/absence data in three male genomes analyzed suggest that it may have retrotransposed carrying a 3' transduced sequence. L1C does not reside within or close to known or predicted genes, so it is not known whether this element is expressed. Although we have no proof that any of the three hot L1s – L1A, L1B, or L1C – is expressed in vivo, four of the hot L1s – three in this study and the disease-producing hot L1, LRE1 – had common alleles that demonstrated highly variable retrotransposition activity in cell culture. These data suggest that the great bulk of hot L1s that are responsible for most in vivo retrotransposition have both hot and cool alleles.

From this large study of alleles of young L1s currently expanding in the human genome, we suggest a model for how full-length L1 insertions (about 35% of the total), evolve in a population (Fig. 5). This work and the fact that nearly all present-day, disease-causing L1 insertions are hot suggest that new insertions in a population are derived nearly exclusively from hot L1s. Later, as a new insertion increases in gene frequency through genetic drift, it also acquires random mutations, some of which reduce its retrotransposition potential from hot to cool, which is the status of the three elements we have studied. As the gene frequency of the L1 increases toward fixation, alleles continue to accumulate mutations that render them either cool or dead for subsequent retrotransposition.

The substantial individual variation in retrotransposition capability that we have found for the three hot elements studied raises an important question: would this degree of individual variation stand if other hot L1s present in the population were included in the analysis? In other words, do L1A, L1B, and L1C combined constitute a significant fraction of the hot L1 activity in world populations? We have used our data and those of Boissinot et al. (2004) to estimate that L1A, L1B, and L1C may account for at least 1/3 of the common hot L1 activity. After other hot L1s are studied in the individuals studied here and in others, we predict that the proportion of individuals at the extremes of the distribution of retrotransposition capability will decrease somewhat, but the difference in retrotransposition potential of individuals at those extremes will increase. Thus, we conclude that individual variation in retrotransposition activity is an important contributor to human genetic diversity.

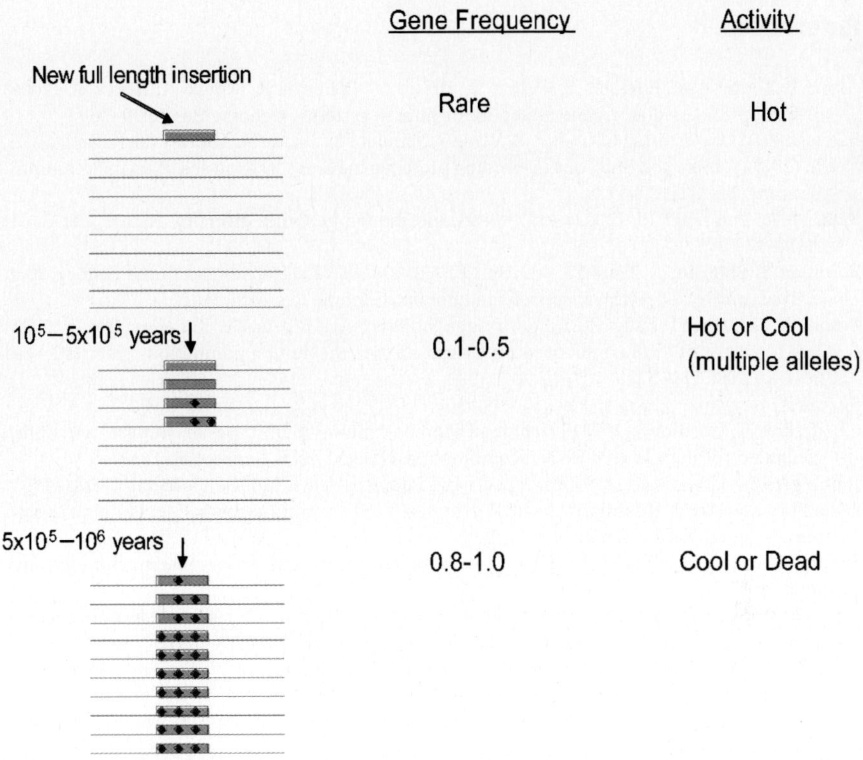

Fig. 5. Model of the evolution of an L1 insertion in a population. Data presented here and evidence that hot L1s account for most new insertions (Brouha et al. 2003) suggest that new insertions are derived from hot L1s. Data on alleles of L1A, L1B, L1C, and LRE1 indicate that, after a hot L1 reaches an intermediate gene frequency in the population, it has a significant proportion of cool alleles. As an L1 approaches fixation, mutations produce cool alleles and dead alleles. *Shaded box* = L1 insertion in chromosomes (*lines*). *Black dots* = mutations

This conclusion has important implications for the role of retrotransposition in producing neurological diversity among human beings. First of all, it is difficult to consider retrotransposition as an important evolutionary mechanism to produce diversity when such a high percentage of the total L1 activity in an individual (at least in cell culture) is contributed by a very small number of very active L1s (a handful or less). That is not to say that L1 elements never had a significant role in human diversity. It is possible that the number of highly active L1s per individual was much greater millions of years ago, and subsequently decreased due to mutation and population bottlenecks. Second, if retrotransposition potential is very low in some individuals and very high in others, it is likely that retrotransposition is an unstable mechanism for effecting change in behavior or learning. Another population bottleneck could eliminate it entirely. We conclude that although retrotransposition is an interesting and exciting possibility for creating diversity, it is unlikely to be a major factor. On the other hand, surprises are what keep biology interesting, and we are prepared to be surprised!

References

Akiva P, Toporik A, Edelheit S, Peretz Y, Diber A, Shemesh R, Novik A, Sorek R (2006) Transcription-mediated gene fusion in the human genome. Genome Res 16:30–36

Babushok DV Ok, Ostertag EM, Chen X, Wang Y, Mandal PK, Okada N, Abrams CS, Kazazian HH, Jr. (2007) A novel testis ubiquitin-binding protein gene arose by exon shuffling in hominoids. Genome Res 17:1129–1138

Batzer MA, Deininger PL (2002) Alu repeats and human genomic diversity. Nature Rev Genet 3:370–379

Boissinot S, Entezam A, Young L, Munson PJ, Furano AV (2004) The insertional history of an active family of L1 retrotransposons in humans. Genome Res 14:1221–1231

Brouha B, Schustak J, Badge RM, Lutz-Prigge S, Farley AH, Moran JV, Kazazian HH Jr (2003) Hot L1s account for the bulk of retrotransposition in the human population. Proc Natl Acad Sci USA 100:5280–5285

Buzdin A, Gogvadze E, Kovalskaya E, Volchkov P, Ustyugova S, Illarionova A, Fushan A, Vinogradova T, Sverdlov E (2003) The human genome contains many types of chimeric retrogenes generated through in vivo RNA recombination. Nucleic Acids Res 31:4385–4390

Deininger PL, Batzer MA (1999) Alu repeats and human disease. Mol Genet Metab 67:183–193

Esnault C, Maestre J, Heidmann T (2000) Human LINE retrotransposons generate processed pseudogenes. Nature Genet 24:363–367

Fanning TG, Singer MF (1987) LINE-1: a mammalian transposable element. Biochim Biophys Acta 910:203–212

Feng Q, Moran JV, Kazazian HH, Jr., Boeke JD (1996) Human L1 retrotransposon encodes a conserved endonuclease required for retrotransposition. Cell 87:905–916

Goodier JL, Ostertag EM, Kazazian HH, Jr. (2000) Transduction of 3'-flanking sequences is common in L1 retrotransposition. Human Mol Genet 9:653–657

Gotea V, Makalowski W (2006) Do transposable elements really contribute to proteomes? Trends Genet 22:260–267

Jurka J (2004) Evolutionary impact of human Alu repetitive elements. Curr Opin Genet Dev 14:603–608

Kazazian HH, Jr. (2004) Mobile elements: drivers of genome evolution. Science 303:1626–1632

Lander ES, Linton LM, Birren B, Nusbaum C, Zody MC, Baldwin J, Devon K, Dewar K, Doyle M, FitzHugh W, Funke R, Gage D, Harris K, Heaford A, Howland J, Kann L, Lehoczky J, LeVine R, McEwan P, McKernan K, Meldrim J, Mesirov JP, Miranda C, Morris W, Naylor J, Raymond C, Rosetti M, Santos R, Sheridan A, Sougnez C, Stange-Thomann N, Stojanovic N, Subramanian A, Wyman D, Rogers J, Sulston J, Ainscough R, Beck S, Bentley D, Burton J, Clee C, Carter N, Coulson A, Deadman R, Deloukas P, Dunham A, Dunham I, Durbin R, French L, Grafham D, Gregory S, Hubbard T, Humphray S, Hunt A, Jones M, Lloyd C, McMurray A, Matthews L, Mercer S, Milne S, Mullikin JC, Mungall A, Plumb R, Ross M, Shownkeen R, Sims S, Waterston RH, Wilson RK, Hillier LW, McPherson JD, Marra MA, Mardis ER, Fulton LA, Chinwalla AT, Pepin KH, Gish WR, Chissoe SL, Wendl MC, Delehaunty KD, Miner TL, Delehaunty A, Kramer JB, Cook LL, Fulton RS, Johnson DL, Minx PJ, Clifton SW, Hawkins T, Branscomb E, Predki P, Richardson P, Wenning S, Slezak T, Doggett N, Cheng JF, Olsen A, Lucas S, Elkin C, Uberbacher E, Frazier M, Gibbs RA, Muzny DM, Scherer SE, Bouck JB, Sodergren EJ, Worley KC, Rives CM, Gorrell JH, Metzker ML, Naylor SL, Kucherlapati RS, Nelson DL, Weinstock GM, Sakaki Y, Fujiyama A, Hattori M, Yada T, Toyoda A, Itoh T, Kawagoe C, Watanabe H, Totoki Y, Taylor T, Weissenbach J, Heilig R, Saurin W, Artiguenave F, Brottier P, Bruls T, Pelletier E, Robert C, Wincker P, Smith DR, Doucette-Stamm L, Rubenfield M, Weinstock K, Lee HM, Dubois J, Rosenthal A, Platzer M, Nyakatura G, Taudien S, Rump A, Yang H, Yu J, Wang J, Huang G, Gu J, Hood L, Rowen L, Madan A, Qin S, Davis RW, Federspiel NA, Abola AP, Proctor MJ, Myers RM, Schmutz J, Dickson M, Grimwood J, Cox DR, Olson MV, Kaul R, Raymond C, Shimizu N, Kawasaki K, Minoshima S,

Evans GA, Athanasiou M, Schultz R, Roe BA, Chen F, Pan H, Ramser J, Lehrach H, Reinhardt R, McCombie WR, de la Bastide M, Dedhia N, Blocker H, Hornischer K, Nordsiek G, Agarwala R, Aravind L, Bailey JA, Bateman A, Batzoglou S, Birney E, Bork P, Brown DG, Burge CB, Cerutti L, Chen HC, Church D, Clamp M, Copley RR, Doerks T, Eddy SR, Eichler EE, Furey TS, Galagan J, Gilbert JG, Harmon C, Hayashizaki Y, Haussler D, Hermjakob H, Hokamp K, Jang W, Johnson LS, Jones TA, Kasif S, Kaspryzk A, Kennedy S, Kent WJ, Kitts P, Koonin EV, Korf I, Kulp D, Lancet D, Lowe TM, McLysaght A, Mikkelsen T, Moran JV, Mulder N, Pollara VJ, Ponting CP, Schuler G, Schultz J, Slater G, Smit AF, Stupka E, Szustakowski J, Thierry-Mieg D, Thierry-Mieg J, Wagner L, Wallis J, Wheeler R, Williams A, Wolf YI, Wolfe KH, Yang SP, Yeh RF, Collins F, Guyer MS, Peterson J, Felsenfeld A, Wetterstrand KA, Patrinos A, Morgan MJ, de Jong P, Catanese JJ, Osoegawa K, Shizuya H, Choi S, Chen YJ; International Human Genome Sequencing Consortium. (2001) Initial sequencing and analysis of the human genome. Nature 409:860–921

Luan DD, Korman MH, Jakubczak JL, Eickbush TH (1993) Reverse transcription of R2Bm RNA is primed by a nick at the chromosomal target site: a mechanism for non-LTR retrotransposition. Cell 72:595–605

Martin SL, Bushman FD (2001) Nucleic acid chaperone activity of the ORF1 protein from the mouse LINE-1 retrotransposon. Mol Cell Biol 21:467–475

Mathias SL, Scott AF, Kazazian HH, Jr., Boeke JD, Gabriel A (1991) Reverse transcriptase encoded by a human transposable element. Science 254:1808–1810

Moran JV, Holmes SE, Naas TP, DeBerardinis RJ, Boeke JD, Kazazian HH, Jr. (1996) High frequency retrotransposition in cultured mammalian cells. Cell 87:917–927

Moran JV, DeBerardinis RJ, Kazazian HH, Jr. (1999) Exon shuffling by L1 retrotransposition. Science 283:1530–1534

Morrish TA, Gilbert N, Myers JS, Vincent BJ, Stamato TD, Taccioli GE, Batzer MA, Moran JV (2002) DNA repair mediated by endonuclease-independent LINE-1 retrotransposition. Nature Genet 31:159–165

Muotri AR, Chu VT, Marchetto MC, Deng W, Moran JV, Gage FH (2005) Somatic mosaicism in neuronal precursor cells mediated by L1 retrotransposition. Nature 435:903–910

Ostertag EM, Kazazian HH, Jr. (2001) Biology of mammalian L1 retrotransposons. Annu Rev Genet 35:501–538

Ostertag EM, Goodier JL, Zhang Y, Kazazian HH, Jr. (2003) SVA elements are nonautonomous retrotransposons that cause disease in humans. Am J Human Genet 73:1444–1451

Pickeral OK, Makalowski W, Boguski MS, Boeke JD (2000) Frequent human genomic DNA transduction driven by LINE-1 retrotransposition. Genome Res 10:411–415

Sayah DM, Sokolskaja E, Berthoux L, Luban J (2004) Cyclophilin A retrotransposition into TRIM5 explains owl monkey resistance to HIV-1. Nature 430:569–573

Scott AF, Schmeckpeper BJ, Abdelrazik M, Comey CT, O'Hara B, Rossiter JP, Cooley T, Heath P, Smith KD, Margolet L (1987) Origin of the human L1 elements: proposed progenitor genes deduced from a consensus DNA sequence. Genomics 1:113–125

Seleme Mdel C, Vetter MR, Cordaux R, Bastone L, Batzer MA, Kazazian HH, Jr. (2006) Extensive individual variation in L1 retrotransposition capability contributes to human genetic diversity. Proc Natl Acad Sci USA 103:6611–6616

Singer MF, Krek V, McMillan JP, Swergold GD, Thayer RE (1993) LINE-1: a human transposable element. Gene 135:183–188

Speek M (2001) Antisense promoter of human L1 retrotransposon drives transcription of adjacent cellular genes. Mol Cell Biol 21:1973–1985

Swergold GD (1990) Identification, characterization, and cell specificity of a human LINE-1 promoter. Mol Cell Biol 10:6718–6729

Wang H, Xing J, Grover D, Hedges DJ, Han K, Walker JA, Batzer MA (2005) SVA elements: a hominid-specific retroposon family. J Mol Biol 354:994–1007

Wei W, Gilbert N, Ooi SL, Lawler JF, Ostertag EM, Kazazian HH, Boeke JD, Moran JV (2001) Human L1 retrotransposition: cis preference versus trans complementation. Mol Cell Biol 21:1429–1439

Zhang Z, Harrison PM, Liu Y, Gerstein M (2003) Millions of years of evolution preserved: a comprehensive catalog of the processed pseudogenes in the human genome. Genome Res 13:2541–2558

Zheng D, Zhang Z, Harrison PM, Karro J, Carriero N, Gerstein M (2005) Integrated pseudogene annotation for human chromosome 22: evidence for transcription. J Mol Biol 349:27–45

From the "RNA World" to Brain Complexity: Generation of Diversity

Alysson R. Muotri[1], *Maria C.N. Marchetto*[1], and *Fred H. Gage*[1]

> "To many neuroscientists one pyramidal cell is just like another.
> I, on the contrary, believe that it is important to
> distinguish the many types (and probably subtypes) of
> pyramidal cells. One can often see that two pyramidal
> cells look quite different."
>
> Francis Crick
> In *The Impact of Molecular Biology on Neuroscience* (1999)

> "I believe there is little reason to question
> the presence of innate systems that are
> able to restructure a genome."
>
> Barbara McClintock
> In *The Dynamic Genome: Barbara McClintocks's Ideas
> in the Century of Genetics* (1992)

Summary

The recent finding that LINE-1 (Long Interspersed Nucleotide Elements-1, or L1) retroelements are active in somatic neuronal progenitor cells has provided a potential additional mechanism for generating neuronal diversification. L1 retrotransposition in the nervous system challenges the idea of static neuronal genomes, adding a new element for neuronal plasticity. Long dismissed as selfish or "junk" DNA, retroelements are thought to be intracellular parasites from our distant evolutionary past. Together with their mutated relatives, retroelement sequences constitute 45% of the mammalian genome, with L1 alone representing 20%. The fact that L1 can retrotranspose in a defined window of neuronal differentiation, changing the genetic information in single neurons in an arbitrary fashion, could allow the brain to develop in distinctly different ways. These characteristics of variety and flexibility may contribute to the uniqueness of an individual brain. However, the extent of the impact of L1 on the neuronal genome is unknown. In this chapter we will discuss the potential influence of L1 retrotransposition during brain development and the evolutionary pressures that may have selected this unexpected machinery of diversity in neuronal precursor cells. The characterization of somatic neuronal diversification will not only be relevant for the understanding of brain complexity and neuronal organization but may also shed

---

[1] Laboratory of Genetics, The Salk Institute for Biological Studies, 10010 North Torrey Pines Road, La Jolla, CA 92037, USA, e-mail: gage@salk.edu

Gage et al.
Retrotransposition, Diversity and the Brain
© Springer-Verlag Berlin Heidelberg 2008

light on the differences in cognitive abilities, personality traits and many psychiatric conditions observed in humans.

Introduction

There are several ways to study brain complexity. Perhaps the broadest view is to analyze the brain's action or its consequences. However, one could take a physiological approach and investigate how different regions of the brain produce a specific task. Others may try to understand the organization and rules of neuronal networks, or how the neural cells are connected to each other through synapses, in a systematic manner. Then, there are the cellular and molecular views. For those views, the ultimate characteristics of a single neuron are present inside the cell and how genes and molecules react to outside stimuli, from the environment or from the interaction with other cells. This reductionist approach will certainly not explain how the brain works, but may provide the necessary tools for understanding how the different levels of organization co-exist to generate perception, action, feelings, memories and thoughts.

It has been known for more than a century, through the work of Camillo Golgi and Santiago Ramon y Cajal, that neurons are specialized cells with a huge diversity of shapes and connections. It is estimated that the human brain contains more than 10 000 different morphological types of neurons. However, neuronal diversity cannot be defined only by morphology or anatomic position. Similar cells, located at the same brain region, may have distinct electrophysiological properties and unique connection within other neurons. Moreover, neurons are extremely plastic cells, allowing extraordinary response upon micro and macro environmental stimulation. Any attempt to understand how the brain works must take into account this huge neuronal diversity. Such diversity is likely the reason why each one of us is unique; even genetically identical twins have different preferences or opinions. But the fundamental mechanisms by which neural stem cells produce such a variety of neuronal types are slowly being revealed.

In contrast to the single mechanism for the production of antibodies (VDJ recombination) in the immune system, several molecular mechanisms contribute to the generation of neuronal diversity (Muotri and Gage 2006). Those mechanisms not only act on the DNA, but also act on the RNA and protein levels, allowing epigenetic modifications to take place. Among these mechanisms, alternative splicing, promoter usage, alternate polyadenylation, RNA editing and post-translation modifications are all part of the genetic tool box present in neuronal precursor cells. However, even such a repertoire is not enough to justify the observed constellation of neuronal types. New mechanisms are likely to be uncovered. We anticipate that novel strategies for the neuronal diversity contribution are hidden in non-coding regions of the genome (Cao et al. 2006; Muotri and Gage 2006). We have recently shown that an engineered human L1 element can retrotranspose in neuronal precursor cells, changing neuron-related gene expression, which, in turn, can influence neuronal cell fate in vitro (Muotri et al. 2005). Moreover, because L1 retrotransposition can also occur in the CNS neuroprogenitor throughout embryo development and in the adult brain of transgenic mice, this unexpected mechanism may contribute to neuronal diversity.

The L1 retrotransposition causes a neuronal genetic mosaicism, i.e., the presence of more than one genetically distinct neuronal type. Such mosaicism might be undetectable unless closely inspected. In fact, genetic mosaicism is frequently overlooked or interpreted as normal variation caused by stochastic developmental factors or the unequal influence of the environment. However, depending on the mosaic nature, frequency, environmental cues, and tissues of origin, even subtle alterations in gene expression can contribute to detectable phenotypic alterations in the organism. Normal processes, such as aging, the generation of immune diversity, and the phenotypic variability between monozygotic twins (such as schizophrenia) can be due to somatic genetic mosaicism (Dipple and McCabe 2000; Machin 1996; Vijg 2000). The stochastic nature of retrotransposon activity, and the large number of genes that this process may affect, could produce an ample spectrum of neuronal diversity, which may affect behavior, cognition and disease risk.

Silencing and Activation of L1 Retrotransposons

L1 retrotransposons can threaten the structure and regulate the expression of the genome in different ways, such as creating new splicing forms, promoter activation, skipping exons or gene inactivation among others (Gilbert et al. 2005; Kazazian 2004). Such a variety of strategies makes L1 retrotransposons the most creative force shaping the genomes throughout evolution.

Deleterious retrotransposition events in the germ line or in early development have resulted in a variety of genetic disorders, and a somatic L1 retrotransposition in man has resulted in a sporadic case of colon cancer (Kazazian 1998; Miki et al. 1992; Ostertag and Kazazian 2001). In plants and other organisms in which transposition is not restricted to the germ line, somatic activity of transposable elements provides the opportunity for a phenotypic variability that can sometimes be stunning with regard to individual genome flexibility (Lisch 2002). In contrast, retrotransposons are often assumed to be silenced in mammalian somatic tissues. This assumption is based on several arguments. First, there is no detectable level of retrotransposon expression in most somatic tissues. However, only a few tissues have been subjected to meticulous analysis, including subtype cell differences. Second, the somatic silencing of L1s fits well in the "selfish" DNA hypothesis, where the mobile elements exist merely to propagate themselves, so there is no reason to transpose in somatic cells. Finally, there is a clear detection bias of somatic retrotransposition, since only visible mutants, usually leading to human diseases, such as cancer, are detectable (Kazazian 2001; Kazazian et al. 1988). The lack of experimental data and the paucity of natural evidence for somatic L1 retrotransposition have led to the view that L1 activity is restricted to early embryonic and germ line cells, suggesting that intrinsic factors may be present or absent in certain cell types responsible for transposition (Mathias and Scott 1993; Prak et al. 2003). Nonetheless, retrotransposon silencing could be physiologically attenuated. DNA methylation is likely the most effective and global strategy against retrotransposon mobility (Yoder et al. 1997). Accordingly, DNA methyltransferase-1 (*Dmnt1*)-deficient mouse embryos have much higher levels of IAP (Intracisternal A-particle) retrotransposon transcripts than their wild-type littermates (Walsh et al. 1998). Repression of retrotransposition is removed under definite conditions during a specific developmental window. One example is the specific induction of IAP elements in the stem cells of the male germ line at undifferentiated stages when they are de-methylated, leading to the hypothesis that a similar mechanism may be found in somatic tissues.

One useful approach to track somatic retrotransposition is the analysis of the L1-EGFP transgenic animal (Muotri et al. 2005; Prak et al. 2003). These mice were engineered to carry an active L1 retrotransposon with an EGFP indicator cassette that only expresses EGFP after retrotransposition and de novo insertions (Fig. 1). Because the assay uses the strong and ubiquitous CMV promoter, it is expected to express EGFP in a large spectrum of somatic cells, if retrotransposition indeed occurs. Obviously, the system will not detect truncated or silenced insertions of the reporter cassette. Of the several somatic tissues analyzed by immunohistochemistry, brain tissue was the only tissue where EGFP expression was detected, specifically in neurons (Muotri et al. 2005). The in vitro cellular assay indicated that L1 retrotransposition actually happened in precursor cells rather than postmitotic neurons. Therefore, neuronal precursor cells might have a greater frequency of L1 retrotransposition than other cell types and/or this finding might be due to the long life of neurons, in contrast to the continuous renewal of other cell types. Either way, the presence of EGFP-positive cells indicates that somatic L1 silencing is incomplete in the brain. This observation suggests that L1 retrotransposons might be activated in neuronal precursor cells and the resultant retrotransposition events could alter the expression of neuronal genes. Such a mechanism could, presumably, generate a large spectrum of genetically distinct neurons, adding to the great neuronal variation that is currently observed in the adult CNS. L1 activation is likely regulated by host factors in equilibrium: too much L1 retrotransposition can cause cell damage and induce the cells to die (Haoudi et al. 2004); too little can limit neuronal diversity. The identification of neuronal host factors responsible for L1 repression and/or activation will be extremely important to understanding how retrotransposition is regulated.

L1 expression is dependent on the activation of its own 5'UTR sequence, which acts as a promoter. The human L1 5'UTR is about 1 kb long, harboring one YY1-binding site that is required for proper transcriptional initiation (Athanikar et al. 2004; Swergold 1990), two Sox (sex determining region of Y-chromosome, SRY, related HMG-box) binding sites (Tchenio et al. 2000) and a runt-domain transcription factor 3 (RUNX3) binding site (Yang et al. 2003). Interestingly, none of these factors is germ-cell specific, suggesting the presence of other, unknown factors. Sox proteins are expressed in a variety of tissues, including neural stem cells (NSCs) and testis (Wegner 1999). The lack of Sox2 allowed activation of neuronal genes and differentiation, suggesting that Sox2 may function as a repressor of differentiation in NSCs (Graham et al. 2003). We demonstrated that a decrease in Sox2 expression during the early stages of neuronal differentiation is correlated with an increase in both L1 transcriptional activity and retrotransposition (Muotri et al. 2005). We propose that L1 retrotransposons are silenced in NSCs due to Sox2-mediated transcriptional repression. Down-regulation of Sox2 accompanies chromatin modifications, such as DNA de-methylation and histone acetylation, which may trigger neuronal differentiation (Fig. 2). Such a mechanism preserves genetic stability in NSCs but allows instability to happen in neuronally committed cells.

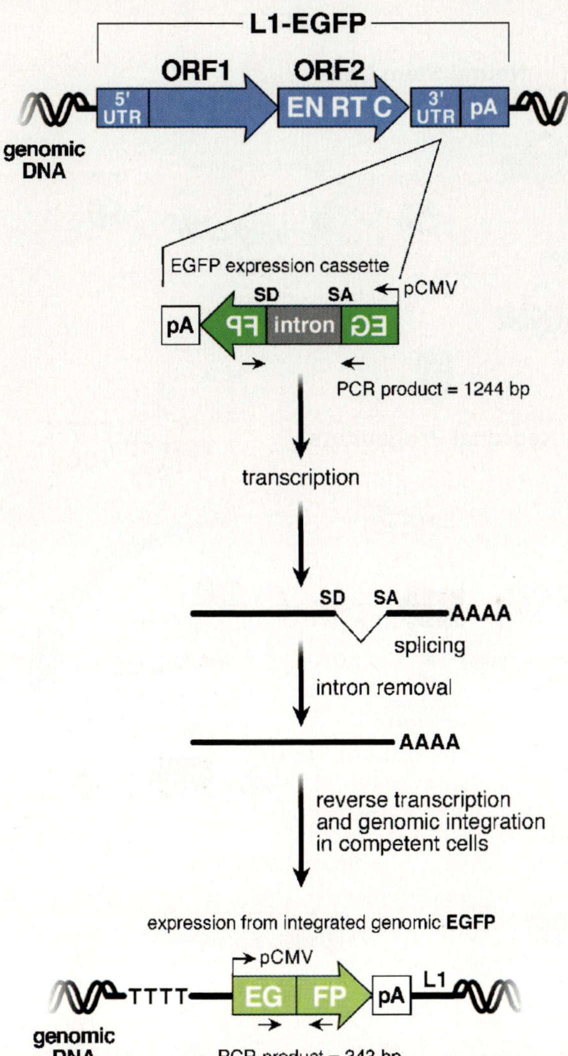

Fig. 1. Detection of L1 retrotransposition in the brains of transgenic mice. The structure of the L1$_{RP}$-EGFP transgene is indicated at the top of the figure. The retrotransposition-competent human L1 (L1$_{RP}$) contains a 5' untranslated region (UTR) that harbors an internal promoter, two open reading frames (ORF1 and ORF2; not drawn to scale), and a 3' UTR that ends in a poly (A) tail. The EGFP retrotransposition indicator cassette consists of a backward copy of the *EGFP* gene whose expression is controlled by the human cytomegalovirus major immediate early promoter (pCMV) and the herpes simplex virus thymidine kinase polyadenylation sequence (pA). This arrangement ensures that EGFP expression will only become activated upon L1 retrotransposition. The *black arrows* indicate PCR primers flanking the intron present in the *EGFP* gene

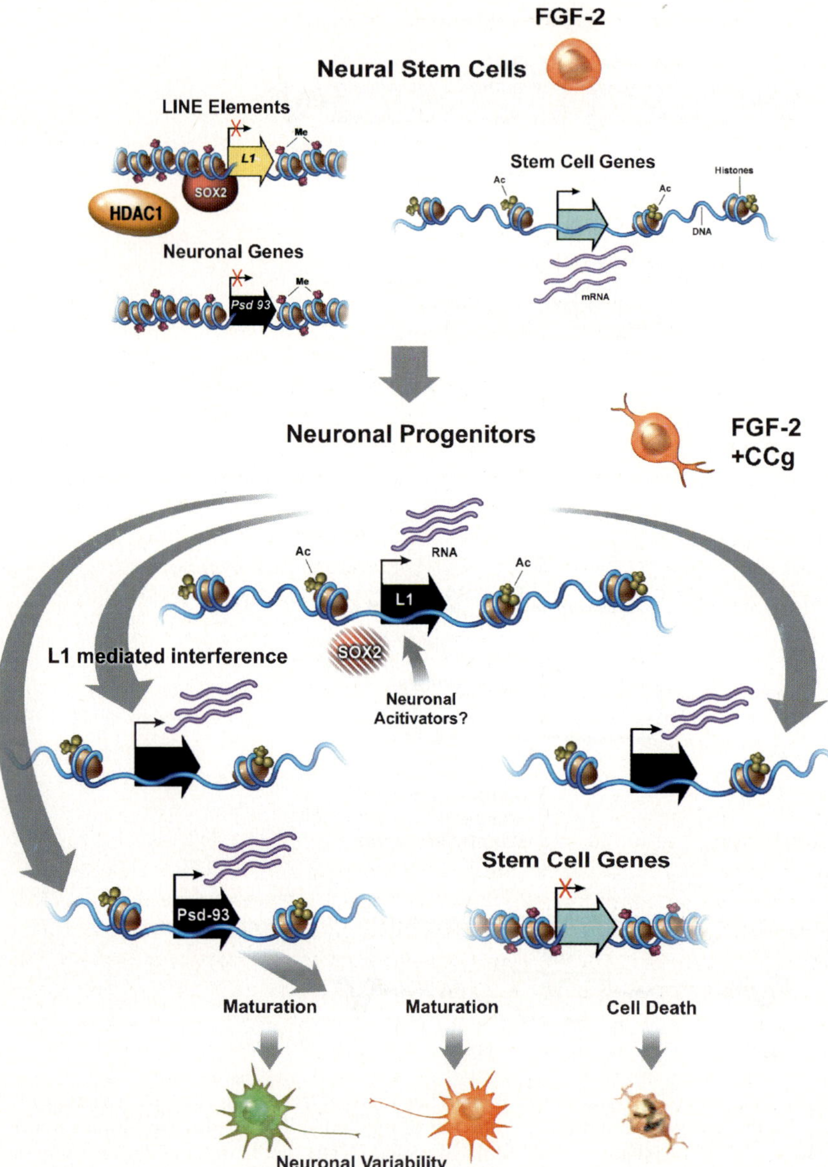

Fig. 2. A model for generation of neuronal diversity by L1 retrotransposition. In neural stem cells, Sox2 expression is correlated with a repression of L1 retrotransposons and neuronal genes. During early phases of neuronal differentiation, there is a reduction in the expression of Sox2 and other neuronal stem cell genes. As a result, L1 transcription can be activated, allowing subsequent retrotransposition into neuronal genes such as for the *Psd-93* gene. The resulting retrotransposition events can alter gene expression, which, in turn, can influence the phenotype of the resulting cell. The functional variability in gene expression induced by L1 retrotransposition could also contribute, in principle, to the high cell death rate observed in adult neurogenesis, where only a few newly born neurons successfully integrate into the pre-existing neuronal network

L1 Targets in Neuronal Progenitor Cells

To cause a significant impact on neuronal genomes, new L1 insertions must target important regulatory regions or genes that are being expressed at the moment of neuroblast differentiation. It is likely that only the combination of multiple L1 events, and not an eventual catastrophic insertion in single neurons, will be ultimately responsible for any change in the neuronal network. But L1 retrotransposition is a dangerous situation for the cell, since L1 insertions can hit essential genes that may induce cell death or even target oncogenes, leading to a neoplastic transformation.

Despite the low number of examples, the sequence data from target insertional sites in rat neuroblasts were often close to or inside neuron-associated genes (Muotri et al. 2005). Even with a small sample, two L1 insertions were located in the same gene, indicating that the integration process might not be completely random. Some of these target genes included an olfactory receptor, ion channel-associated genes and a cadherin receptor (Muotri et al. 2005). An L1 insertion in the promoter region of the *Psd-93* gene, encoding a post-synaptic density protein involved in different aspect of synapse formation, significantly increased gene expression level and, consequently, accelerated neuronal maturation in culture. Despite the fact that randomness seems to be the best way for L1 to survive during evolution when they are active in germ cells, somatic insertions might be controlled by local microvariations in DNA chromatin structure that depend on different host factors in specific subsets of cell types. Thus, we propose that L1 insertions in the nervous system are somehow guided to specific gene targets. In a similar way, the yeast Ty1 transposon is highly nonrandom in vivo, being preferentially inserted upstream of tRNA genes (Bachman et al. 2004; Devine and Boeke 1996).

In the L1-EGFP transgenic mice, we followed the retrotransposition of a single human L1 element and retrotransposition was detected by EGFP expression. However, the indicator cassette did not reflect a direct measurement of the 3 000 estimated endogenous active L1 retrotransposons (DeBerardinis et al. 1998; Goodier et al. 2001). Moreover, as pointed out before, the L1 retrotransposition assay did not report EGFP-truncated or silenced insertions. Additionally, it certainly did not account for the indirect, in trans, L1-mediated insertions of Alus, retrotransposition-defective L1s, and other non-autonomous RNAs. Virtually any RNA molecule can be subject to retrotransposition if hijacked by L1 machinery. Thus, every single developing neuron can potentially carry L1-mediated events, and if part of the resultant insertions occurs in genes expressed during neuronal development, altering gene expression, then it is possible that brain development could be significantly affected by L1 retrotransposition. It has been proposed that stochastic gene expression might be a fundamental part of development and differentiation and, where it is advantageous, these stochastic patterns are retained in the adult organism (Fiering et al. 2000). We speculate that these new L1 retrotransposition events are stably integrated into the genomes of individual neurons during the entire life of the organism. These insertions then act in a stochastic fashion, working as "controlling elements," fine-tuning to increase the probability that genes will be differentially transcribed. The model is consistent with neuroblast differentiation, in which similar cells are subjected to the same environmental stimuli but do not respond uniformly. Thus, new insertions in neurons represent genomic "scars" that may have the potential to influence the fate of the resultant cells and, consequently, the function of the neuronal network.

The study of the human L1 5'UTR promoter during neuronal differentiation revealed that L1 activation occurs in the initial stages of cell differentiation. That is exactly the same time that several neuronal genes, such as *NeuroD1*, are upregulated and several cell cycle genes are downregulated (Hsieh et al. 2004; Zhao and Gage 2002). Additionally, the strong anti-mitotic small modulatory NRSE dsRNA, responsible for the neuronal fate of NSCs, is expressed in initial steps of differentiation, activating several NRSE-containing neuron-specific genes and stopping the cell cycle (Kuwabara et al. 2004). These data suggest that there is an orchestrated regulation during neuronal differentiation, avoiding an eventual cell transformation. Such an idea conforms with the low incidence of neuroblastomas (Zhu and Parada 2002) but does not exclude the possibility that an abnormal L1 retrotransposition might lead to a neoplastic transformation in CNS cells.

Taken together, a specific regulation of L1 retrotransposon activity that takes into account its "non-random" neuronal insertion and a specific window of time during cell differentiation may turn a potentially harmful phenomenon into a useful one. The problem now, as with most novel scientific debates, is one of quantification and significance. Future technologies for single-cell endogenous L1 activity assays will bring new insights into the problem. Moreover, the generation of three-dimensional brain mapping depicting the occurrence of L1 retrotransposition will allow the visualization of preferential target neuronal subtypes. The comparison of normal brains with brains where L1 activity is misregulated will provide the structural organization for the design of algorithms that predict eventual retrotransposition-affected neuronal networks or systems.

Evolutionary Consequences of L1 Impact in Neuronal Genomes

One of the most remarkable findings from the sequencing of the human genome is that retrotransposable elements make up a significant portion of the human DNA (Deininger et al. 2003). Based on reverse transcriptase (RT) phylogeny, L1 elements are most closely related to the group II introns of mitochondria and eubacteria (Cavalier-Smith 1991; Xiong and Eickbush 1990). These studies revealed that the RT enzyme is extremely old and that retroelements can be viewed as relics or molecular fossils of the first primitive replication systems in the progenote. The origin of retroelements possibly traces back to the conversion of RNA-based systems, the "RNA World" (Orgel, 2004), to modern "DNA-based" systems. Current models suggest that these mobile introns of eubacteria were transmitted to eukaryotes during the initial fusion of the eubacterial and archaebacterial genomes or during the symbiosis that gave rise to the mitochondria, generating the modern-day spliceosomal introns (Zimmerly et al. 1995). Further acquisition of an endonuclease enzyme and a promoter sequence certainly represented important steps in the evolution of L1 retrotransposons, providing autonomy for L1s to insert into many locations throughout the genome.

The apparent lack of obvious function of retroelements in the genome suggests that transposable elements are "selfish DNA," acting as parasites in the genome to propagate themselves. This idea has long puzzled scientists and inspired the concept of "junk DNA" to illustrate the idea that such sequences were mere evolutionary remnants (Doolittle and Sapienza 1980; Orgel and Crick 1980). However, the recognition that retrotransposons can actively reshape the genome is slowly challenging this terminology. Moreover, the mammalian genome has suffered waves of transposon bombardment, but the constant, single lineage of L1 history reveals that active L1 were never absent from mammals' genomes during evolution, suggesting an inextricable link between L1 and their hosts (Furano et al. 2004). The relationship between transposons and their hosts is probably not entirely antagonistic, as several host genes have a high degree of homology to one or more transposable elements. Evidence in the literature points to a somatic function for L1 transcripts, involving cell proliferation (Kuo et al. 1998), differentiation (Mangiacasale et al. 2003) and early embryo development (Pittoggi et al. 2003). Moreover, it is difficult to reconcile why the genome would need so many copies of retrotransposons and whether this expansion has any correlation with retrotransposition itself. The restricted activity of retrotransposons in germ or early embryonic cells apparently fits well with the "selfish DNA" concept, since new insertions will be passed to the next generations, but somatic insertions pose a conundrum. According to the symbiotic theory, it is advantageous to any transposable element to promote host mating, securing the propagation of the "master" elements to the next generations. From this perspective, it is not surprising that advantageous insertional events in the brain, resulting in the better (cultural and social) fitness of the individual organism, also can contribute to the host mating.

The evolution of the CNS provided a notable selective advantage, as information about the environment could be processed rapidly and would allow organisms to more readily meet the challenges of ever-changing environmental conditions. Moreover, epigenetic modification allowed the non-genetic transfer of information or transmission of "culture" at an unprecedented magnitude. Such specialization is highly dependent on the cognitive levels acquired by the species that are directly linked to the complexity of the neuronal network. Therefore, the advantages gained by retaining the mechanisms for somatic retrotransposition may outweigh the cost of a less plastic nervous system. In fact, such a strategy expands the number of functionally distinct neurons that could be produced from a given stem cell gene pool (Muotri and Gage 2006). This characteristic of variety and flexibility may contribute to the uniqueness of an individual brain, even between genetically identical twins. Mobile elements in the brain may be part of the conserved core process responsible for evoking the facilitated, complex, non-random phenotypical variation on which selection may act. It is remarkable to imagine that the brain is a consequence of ancient retrotransposition in eukaryotic cells. Such a possibility has not been considered before, but it was suggested to us by the experimental results.

The identification of L1 elements as potential creative somatic shapers of transcriptional complexity in neuronal genomes may be an important phenomenon for developmental neurosciences. The hypothesis that L1 activity is responsible for "fine-tuning" neuronal wiring waves requires the merger of different fields and may consequently open new ways of considering individual differences and the neuronal correlates of human cognition. Rigorous experimental proof of this model will require attenuation of retrotransposition activity from the mammalian genome and comparing their behavior to that of wild type animals. Nonetheless, the experimental approach presents a major methodological challenge for molecular biologists, since a canonical single-gene knockout strategy is no longer suitable. On the other hand, the study of abnormal activation of L1 retrotransposition in the brain may elucidate complex neurological syndromes, permitting an understanding of diseases at a different level.

Acknowledgements. A.R.M is supported by the Rett Syndrome Research Foundation, M.C.N.M. is supported by the George E. Hewitt Foundation for Medical Research and F.H.G. is supported by the Lookout Fund and the National Institutes of Health: National Institute on Aging and National Institute of Neurological Disease and Stroke.

References

Athanikar JN, Badge RM, Moran JV (2004) A YY1-binding site is required for accurate human LINE-1 transcription initiation. Nucleic Acids Res 32:3846–3855

Bachman N, Eby Y, Boeke JD (2004) Local definition of Ty1 target preference by long terminal repeats and clustered tRNA genes. Genome Res 14:1232–1247

Cao X, Yeo G, Muotri AR, Kuwabara T, Gage FH (2006) Noncoding RNAs in the mammalian central nervous system. Annu Rev Neurosci Feb. 15:78–103

Cavalier-Smith T (1991) Intron phylogeny: a new hypothesis. Trends Genet 7:145–148

DeBerardinis RJ, Goodier JL, Ostertag EM, Kazazian HH Jr (1998) Rapid amplification of a retro-transposon subfamily is evolving the mouse genome. Nature Genet 20:288–290

Deininger PL, Moran JV, Batzer MA, Kazazian HH Jr (2003) Mobile elements and mammalian genome evolution. Curr Opin Genet Dev 13:651–658

Devine SE, Boeke JD (1996) Integration of the yeast retrotransposon Ty1 is targeted to regions upstream of genes transcribed by RNA polymerase III. Genes Dev 10:620–633

Dipple KM, McCabe ER (2000) Phenotypes of patients with "simple" Mendelian disorders are complex traits: thresholds, modifiers, and systems dynamics. Am J Human Genet 66:1729–1735

Doolittle WF, Sapienza C (1980) Selfish genes, the phenotype paradigm and genome evolution. Nature 284:601–603

Fiering S, Whitelaw E, Martin DI (2000) To be or not to be active: the stochastic nature of enhancer action. Bioessays 2:, 381–387

Furano AV, Duvernell DD, Boissinot S (2004). L1 (LINE-1) retrotransposon diversity differs dramatically between mammals and fish. Trends Genet 20:9–14

Gilbert N, Lutz S, Morrish TA, Moran JV (2005). Multiple fates of L1 retrotransposition interme-diates in cultured human cells. Mol Cell Biol 25:7780–7795

Goodier JL, Ostertag EM, Du K, Kazazian HH Jr (2001). A novel active L1 retrotransposon subfamily in the mouse. Genome Res 11:1677–1685

Graham V, Khudyakov J, Ellis P, Pevny L (2003) SOX2 functions to maintain neural progenitor identity. Neuron 39:749–765

Haoudi A, Semmes OJ, Mason JM, Cannon RE (2004) Retrotransposition-competent human LINE-1 induces apoptosis in cancer cells with intact p53. J Biomed Biotechnol 2004:185–194

Hsieh J, Nakashima K, Kuwabara T, Mejia E, Gage FH (2004) Histone deacetylase inhibition-mediated neuronal differentiation of multipotent adult neural progenitor cells. Proc Natl Acad Sci USA 101:16659–16664

Kazazian HH Jr (1998) Mobile elements and disease. Curr Opin Genet Dev 8:343–350

Kazazian HH Jr (2001) Retrotransposon insertions in germ cells and somatic cells. Dev Biol (Basel) 106:307–313; discussion 313–304, 317–329

Kazazian HH Jr (2004) Mobile elements: drivers of genome evolution. Science 303:1626–1632

Kazazian HH Jr, Wong C, Youssoufian H, Scott AF, Phillips DG, Antonarakis SE (1988) Haemophilia A resulting from de novo insertion of L1 sequences represents a novel mechanism for mutation in man. Nature 332:164–166

Kuo KW, Sheu HM, Huang YS, Leung WC (1998) Expression of transposon LINE-1 is relatively human-specific and function of the transcripts may be proliferation-essential. Biochem Biophys Res Commun 253:566–570

Kuwabara T, Hsieh J, Nakashima K, Taira K, Gage FH (2004) A small modulatory dsRNA specifies the fate of adult neural stem cells. Cell 116:779–793

Lisch D (2002) Mutator transposons. Trends Plant Sci 7:498–504

Machin GA (1996) Some causes of genotypic and phenotypic discordance in monozygotic twin pairs. Am J Med Genet 61:216–228

Mangiacasale R, Pittoggi C, Sciamanna I, Careddu A, Mattei E, Lorenzini R, Travaglini L, Landriscina M, Barone C, Nervi C, Lavia P, Spadafora C (2003) Exposure of normal and transformed cells to nevirapine, a reverse transcriptase inhibitor, reduces cell growth and promotes differentiation. Oncogene 22:2750–2761

Mathias SL, Scott AF (1993) Promoter binding proteins of an active human L1 retrotransposon. Biochem Biophys Res Commun 191:625–632

Miki Y, Nishisho I, Horii A, Miyoshi Y, Utsunomiya J, Kinzler KW, Vogelstein B, Nakamura Y (1992) Disruption of the APC gene by a retrotransposal insertion of L1 sequence in a colon cancer. Cancer Res 52:643–645

Muotri AR, Chu VT, Marchetto MC, Deng W, Moran JV, Gage FH (2005) Somatic mosaicism in neuronal precursor cells mediated by L1 retrotransposition. Nature 435:903–910

Muotri AR, Gage FH (2006) Generation of neuronal variability and complexity. Nature 441:1087–1093

Orgel LE (2004) Prebiotic chemistry and the origin of the RNA world. Crit Rev Biochem Mol Biol 39:99–123

Orgel LE, Crick FH (1980) Selfish DNA: the ultimate parasite. Nature 284:604–607

Ostertag EM, Kazazian HH Jr (2001) Biology of mammalian L1 retrotransposons. Annu Rev Genet 35:501–538

Pittoggi C, Sciamanna I, Mattei E, Beraldi R, Lobascio AM, Mai A, Quaglia MG, Lorenzini R, Spadafora C (2003) Role of endogenous reverse transcriptase in murine early embryo development. Mol Reprod Dev 66:225–236

Prak ET, Dodson AW, Farkash EA, Kazazian HHJr (2003) Tracking an embryonic L1 retrotransposition event. Proc Natl Acad Sci USA 100:1832–1837

Swergold GD (1990) Identification, characterization, and cell specificity of a human LINE-1 promoter. Mol Cell Biol 10:6718–6729

Tchenio T, Casella JF, Heidmann T (2000) Members of the SRY family regulate the human LINE retrotransposons. Nucleic Acids Res 28:411–415

Vijg J (2000) Somatic mutations and aging: a re-evaluation. Mutat Res 447:117–135

Walsh CP, Chaillet JR, Bestor TH (1998) Transcription of IAP endogenous retroviruses is constrained by cytosine methylation. Nature Genet 20:116–117

Wegner M (1999) From head to toes: the multiple facets of Sox proteins. Nucleic Acids Res 27:1409–1420

Xiong Y, Eickbush TH (1990) Origin and evolution of retroelements based upon their reverse transcriptase sequences. Embo J 9:3353–3362

Yang N, Zhang L, Zhang Y, Kazazian HH Jr (2003) An important role for RUNX3 in human L1 transcription and retrotransposition. Nucleic Acids Res 31:4929–4940

Yoder JA, Walsh CP, Bestor TH (1997) Cytosine methylation and the ecology of intragenomic parasites. Trends Genet 13:335–340
Zhao X, Gage FH (2002) Expressing the central nervous system. Neurochem Res 27:953–954
Zhu Y, Parada LF (2002) The molecular and genetic basis of neurological tumours. Nature Rev Cancer 2, 616–626
Zimmerly S, Guo H, Perlman PS, Lambowitz AM (1995) Group II intron mobility occurs by target DNA-primed reverse transcription. Cell 82:545–554

Endogenous Retroviruses
and Human Neuropsychiatric Disorders

Robert H Yolken[2], *Håkan Karlsson*[1], *Ioannis Bossis*[2], *Linnéa Asp*[1], *Faith Dickerson*[3], *Christoffer Nellåker*[1], *Michael Elashoff*[4], *Elizabeth Rubalcaba*[2], and *Raphael P. Viscidi*[2]

Summary

Schizophrenia and related disorders are devastating human neuropsychiatric disorders of complex etiology. Epidemiological and family studies indicate both genetic and environmental contributions to disease etiology. We propose that the altered expression of endogenous retroviruses such as HERV-W contribute to some cases of schizophrenia. We present both theoretical considerations and experimental evidence supporting this association. The further study of endogenous retroviral expression within the central nervous system might lead to new methods for the diagnosis and treatment of schizophrenia and related disorders.

Introduction

Schizophrenia is a pervasive neuropsychiatric disorder with worldwide prevalence. The characteristic features of schizophrenia are "positive" symptoms of hallucinations or delusions that are often accompanied by "negative" symptoms, such as emotional withdrawal or amotivation. There is a great deal of individual variation in the symptoms of schizophrenia, its clinical course, and response to medication. Pharmacological treatments may reduce the severity of symptoms, but their effects are typically incomplete. The effects of schizophrenia are devastating for the affected individuals, their families, and society at large. In most areas of the world, schizophrenia is a major cause of premature death, homelessness, and suicide (Jablensky 2000). Because schizophrenia generally has the onset of symptoms in early adulthood and persists for life, it is associated with a high burden of disease (Rossler et al. 2005). According to the World Health Organization, schizophrenia is the eighth leading cause of lost life years worldwide in individuals 15–44 years of age (WHO World Health Report 2001). Schizophrenia is also associated with a number of non-psychiatric medical conditions, such as heart disease (Davidson 2002), diabetes (Zimmet 2005), metabolic syndrome (Meyer et al. 2005), and other autoimmune disorders (Eaton et al. 2006). Schizophrenia also shares epidemiologic and phenotypic features with other serious neuropsychiatric disorders,

[1] Department of Neuroscience, Karolinska Institutet, Retzius v 8, 17177 Stockholm, Sweden
[2] Stanley Laboratory of Developmental Neurovirology, Johns Hopkins University School of Medicine, Baltimore, MD, USA, e-mail: yolken@mail.jhmi.edu
[3] Sheppard Pratt Health System, Baltimore, MD, USA
[4] Stanley Medical Research Institute, Chevy Chase, MD, USA

Gage et al.
Retrotransposition, Diversity and the Brain
© Springer-Verlag Berlin Heidelberg 2008

including bipolar disorder (Maier et al. 2005; Moller 2003). There are currently no reliable laboratory diagnostic assays for the diagnosis of schizophrenia, the determination of disease progression or the monitoring of response to medications. Consequently, there is a pressing need for an increased understanding of the etiopathogenesis of schizophrenia and for the development of new methods for diagnosis and therapy.

Current evidence suggests that schizophrenia is the result of both genetic and environmental factors. Family and adoption studies have indicated a strong genetic contribution to schizophrenia risk. Individuals who have a first-degree relative with schizophrenia have an approximately 10-fold increased risk of schizophrenia as compared to the general population. Individuals who have a monozygotic twin with schizophrenia have an approximately 50-fold increased risk (Maki et al. 2005). However, it is notable that many individuals with schizophrenia do not have a first-degree relative with schizophrenia and that, on a population basis, having a first-degree relative with schizophrenia does not contribute greatly to the attributable risk for this disorder (Mortensen et al. 1999).

Based on the familial association with schizophrenia, a large number of studies have attempted to find disease-associated genes. These studies have employed a wide variety of genetic techniques, including family-based linkage analysis (Hovatti et al. 1998), transmission disequilibrium (Li et al. 2001), sib-pair analysis (Schwab et al. 2005), and case-control studies of single nucleotide polymorphisms (Chen et al. 2004). Such studies have identified a large number of genetic regions and specific genes that appear to be associated with increased risk of schizophrenia in some populations. However, to date no genes of major effect (relative risk or odds ratios (>2.0)) have been found for schizophrenia or the related disorders (Delisi and Fleischhaker 2007). The genes that have been found generally have lower relative risks, are inconsistent among populations, and are shared among psychiatric diagnoses (Gilbody et al. 2007). Attempts to correlate results in different populations have also been hampered by complex haplotype structure leading, in some cases, to the overestimation of the concordance of genetic findings across populations (Mutsuddi 2006).

Because of the limitations of genetic studies, there has also been interest in the identification of environmental factors that might modify gene expression. The role of environmental factors is consistent with epidemiological studies indicating variation in the rate of schizophrenia in terms of season of birth, year of birth, urban birth, and location within a relatively small geographic area (Kirkbride et al. 2006). Increased risk of schizophrenia has also been associated with complications of pregnancy, particularly pre-eclampsia, which has been associated with a 2.5-fold risk of increased schizophrenia in the offspring (Dalman et al. 1999). It is of note that these epidemiological factors point to events occurring early in life. These exposures are thus consistent with the concept that changes in the brain in early life may be associated with abnormalities occurring in adolescence or young adulthood. The reasons for the expression of the symptoms of schizophrenia in late adolescence or early adulthood despite exposures in early life are not known with certainty, but they might be related to neurodevelopmental changes in brain structure occurring during childhood and adolescence (Lewis and Levitt 2002), the effects of hormonal activation (Stevens 2002), the long-term consequences of the immune response to infection (Wright et al. 1993) or the reactivation of latent infectious agents (Torrey 1988).

While there are a number of environmental factors that might contribute to the risk of schizophrenia, most research has been directed at the possible role of infectious agents. In light of the above findings, studies of specific infections have focused on ones which might occur in pregnancy or early life. Specific infectious agents that have been associated with increased risk of schizophrenia following exposure in early life include *Toxoplasma gondii* (Mortensen et al. 2006), herpes simplex virus type 2 (Buka et al. 2001), rubella virus (Brown and Susser 2002), influenza viruses (Brown et al. 2004), enteroviruses (Rantakallio et al. 1997), and agents that cause bacterial meningitis (Abrahao et al. 2005). In addition, increased levels of antibodies to infectious agents have been found in adults with schizophrenia as compared to controls in different populations. These agents include *Toxoplasma gondii*, cytomegalovirus, Epstein Barr virus (Behr et al. 2006; Delisi et al. 1986; Leweke et al. 2004) Borna disease virus (Bode et al. 2001) and Borrelia burgdorferi (Fallon and Nields 1994). In addition, DNA from chlamydial organisms has been found in the blood of individuals with schizophrenia in increased levels as compared to controls (Fellerhoff et al. 2007). Herpes simplex virus Type 1, while not associated with an increased risk of schizophrenia, has been associated with cognitive impairment and structural alterations in the brains of individuals with schizophrenia (Dickerson et al. 2003; Prasad et al. 2007).

Numerous other agents have been associated with acute psychotic events similar to those that occur in individuals with schizophrenia. These include human immunodeficiency virus type 1 (Koutsilieri et al. 2002; Atlas et al. 2007), tick-borne encephalitis virus (Chlabicz et al. 1996), hepatitis B virus (Weber et al. 1994), *Taenia solium* (Meza et al. 2005), plasmodia (Tilluckdharry et al. 1996; Alao and Dewan 2001) and other parasitic agents (Singh and Singh 2000) capable of infecting the central nervous system (CNS).

The most consistent association between a specific infectious agent and risk of schizophrenia has been with the apicomplexan organism, *Toxoplasma gondii*. A recent meta-analysis of 23 independent studies from different areas of the world found that serological evidence of Toxoplasma infection is associated with an increased risk of schizophrenia with an odds ratio of approximately 2.7 (Torrey et al. 2007). Toxoplasma antibodies in pregnancy have also been associated with increased risk of schizophrenia in later life in two independent studies (Mortensen et al. 2006; Brown et al. 2005). It is of note that Toxoplasma infection also causes altered behavior in animal model systems and that this behavior can be normalized by treatment with anti-Toxoplasma medications (Webster et al. 2006).

Increased levels of cytomegalovirus infection have also been found to be associated with schizophrenia in several different adult populations (Leweke et al. 2004). Cytomegalovirus DNA has also been identified in the brain tissue of a small number of individuals with schizophrenia (Moises et al. 1988). The associations between the other infectious agents and the risk of schizophrenia have been more variable with relatively low odds ratios and differences across populations.

Endogenous Retroviruses and Schizophrenia: Theoretical Considerations

Human endogenous retroviruses (HERV) are assumed to be remnants of ancient retroviral infections of our ancestors' germ-line cells. HERV sequences constitute approximately 3–8% of the human genome and can be classified into at least 31 families. Tissue-specific hybridization patterns to arrays of sequences representative of different HERV families was recently reported, indicating a discrete and diversified regulation of their transcriptional activities. Based on the evolving understanding of both endogenous retroviruses and schizophrenia, we have postulated that endogenous retroviruses contribute to the etiopathogenesis of some cases of this disorder. The specific hypotheses are as follows:

1. Some cases of schizophrenia are related to the aberrant expression of endogenous retroviruses within the CNS.
2. Individual differences in the response to endogenous retrovirus expression are related to
   a) polymorphisms within the endogenous retroviruses and adjacent genomic sequences;
   b) exposure to infectious agents and other environmental factors relating to endogenous retrovirus interaction in utero and in later life; and
   c) genetic variation in small molecule transport molecules within the CNS, some of which also function as retrovirus receptors.

Findings relating to the epidemiology and pathogenesis consistent with a possible role for endogenous retroviruses include the following:

1. the widespread distribution of endogenous retroviruses in the human genome is consistent with the linkage of schizophrenia risk to many different genomic regions and complex haplotype structure (Mutsuddi et al. 2006; Delisi and Fleischhaker 2007);
2. the fact that endogenous retroviruses can be activated by infecting viruses or protozoa is consistent with the role of these infectious agents in the pathogenesis of schizophrenia, as discussed above. Specifically, endogenous retroviruses have been shown to be activated by *Toxoplasma gondii* (Frank et al. 2006) and cytomegaloviruses (Nelson et al. 1999), as well as by other human herpes viruses (Christensen et al. 2007; Hsiao et al. 2006);
3. aberrant expression of human endogenous retroviruses has been associated with autoimmune disorders (Anthony et al. 2006; Colmegna and Garry 2006), which are also common in individuals with schizophrenia (Strous and Shoenfeld 2006);
4. human endogenous retroviruses are widely expressed in the human placenta during fetal development, a critical time for the action of environmental factors in schizophrenia (Malassine et al. 2005; Rote et al. 2004). Of particular interest is the possible relationship between HERV expression and preeclampsia (Lee et al. 2001), in light of the relationship between preeclampsia and subsequent risk of schizophrenia;
5. many endogenous retroviruses are active in the human CNS (Weis et al. 2007a) and are differentially expressed in neuroinflammatory disorders (Anthony et al. 2006; Mameli et al. 2007); and
6. some endogenous retroviruses employ neuroactive molecules as receptors. For example, the receptor for HERV-W is the ASCT1/ASCT2 family of neutral amino acid transporters, which are responsible for the transport of excitatory amino acids within the CNS (Marin et al. 2003; Lavillette et al. 2002). Levels of ASCT1 have been

found to differ in the brains of individuals with schizophrenia as compared to controls (Weis et al. 2007b).

It is of note that long interspersed nuclear elements (LINEs) and other retroelements fulfill some, but not all, of these criteria relating to an elevated risk of developing schizophrenia.

This review will present both previously published and new data relating to the potential role of endogenous retroviruses in the etiopathogenesis of schizophrenia and related disorders. This review will focus largely on the HERV-W family of retroviruses, since this group has been the subject of the largest number of studies relating to schizophrenia etiopathogenesis.

Altered HERV-W Expression in Schizophrenia

Using degenerate primers directed towards the retroviral *pol* gene (Tuke et al. 1997), we investigated cerebrospinal fluids (CSFs) obtained from patients during their first hospitalization for schizophrenia or schizoaffective disorder for the presence of retroviral RNA. Approximately one third of the samples tested positive by polymerase chain reaction (PCR). Sequencing of each of these products verified that they were of retroviral origin and similar to endogenous retroviruses present in the human genome (HERV; Karlsson et al. 2001). The majority of the sequences showed a high degree of similarity to sequences previously identified in the CSF of patients suffering from multiple sclerosis (MS), denoted MS-associated retrovirus (MSRV; Perron et al. 1997). Probes derived from MSRV sequences were subsequently used to identify a novel family of HERV denoted HERV-W (Blond et al. 1999). Sequences similar to HERV-W were reported to be differentially present in the serum of MS patients (Garson et al. 1998) as well as in plasma from recent-onset schizophrenia patients (Karlsson et al. 2004) as compared to control individuals. More recently, Huang and coworkers (2006) detected retroviral RNA, as well as antibody reactivity, in one third of patients with recent-onset schizophrenia but not in control individuals. These sequences were also of endogenous origin with the highest reported similarity to ERV9.

Based on these findings, we have directed additional studies toward the analysis of HERV-W. According to Pavlicek et al. (2002), the human genome contains in excess of 600 HERV-W elements, the majority of which are long terminal repeat regions (LTR) lacking internal sequence (*gag, pol* and *env* genes). The remaining elements have been classified into two major categories: a total of 77 retroelements with proviral structure containing intact LTRs and complete or partial internal sequences; and 149 pseudoelements with internal sequences, lacking the regulatory U3 region of the 5'-LTR and the U5 region of the 3'-LTR. Structurally these copies resemble retroviral mRNAs and are thought to originate from LINE-mediated reverse transcription of such mRNAs. The remaining elements were grouped together in a third category based on the absence of the diagnostic regions (Pavlicek et al. 2002). Since these latter groups lack regulatory promoter regions, they were suggested to be non-transcribed (Costas 2002; Pavlicek et al. 2002). Whether this is actually the case has, however, not been investigated. Indeed, very little information exists regarding the expression and transcriptional control of individual HERV-W elements. This gap in knowledge is in part explained by the notion that such non-coding RNAs represent transcriptional noise and therefore lack biological relevance but also by the methodological challenges associated with studies of transcripts from large numbers of closely related sequences. Although large-scale studies of transcripts from the different HERV families have been conducted by means of hybridization to arrays of representative sequences (Seifarth et al. 2005) or by use of degenerate primers and probes in real-time PCR assays (Forsman et al. 2005), the individual elements are not resolved by such methods.

Identification of the Source of HERV-W Transcripts

In a recent report, we used sequence-dependent variations in melting temperatures of double-stranded real-time PCR products for studying expression of HERV-W elements (Nellaker et al. 2006). This approach has recently been employed for strain identification in clinical and veterinary virology (Pham et al. 2005; Waku-Kouomou et al. 2006), typing of mycoplasma strains (Harasawa et al. 2005), genotyping of HLA variants (Graziano et al. 2005), detection of translocations in cancers (Bohling et al. 1999) and scanning for single nucleotide polymorphisms (Germer and Higuchi 1999).

By this approach, in combination with extensive sequencing of the different products, we reported cell line-specific transcription patterns of genomic regions harboring HERV-W elements during base-line conditions. We also reported that not only herpes simplex type 1 virus, but also influenza A virus infection of different human cell lines results in significantly increased levels of such transcripts. In each of the different cell-lines, the virus infection induced a unique response in terms of transactivated elements (see Fig. 1). Mapping of these elements indicated a basal and regulated expression of not only elements with a proviral structure but also elements lacking the regulatory

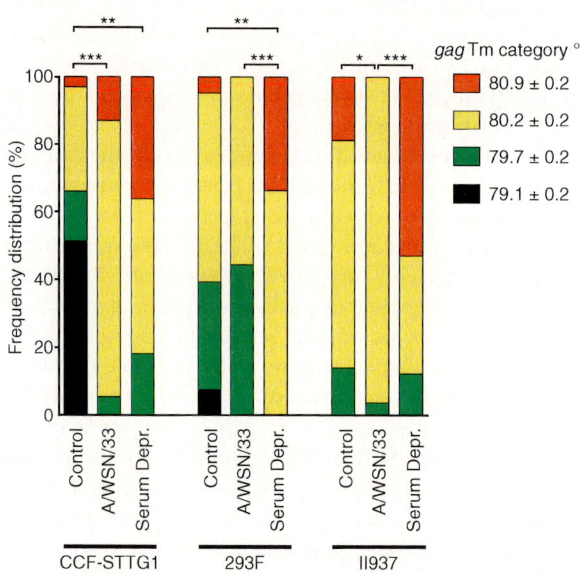

Fig. 1. Frequency distribution plot of melting temperatures representing variations in sequence of amplified HERV W *gag* sequences. Sequences amplified from the cell lines investigated; CCF-STTG1, 293F and U937 were compared between control cells and cells exposed to influenza A/WSN/33 virus or serum deprivation. Temperature ranges indicate discrete temperature categories representing distinct sequences

LTR regions. Serum deprivation induced a similar, but not identical, response in all the cell lines investigated. Taken together, these findings suggest the existence of intrinsic mechanisms governing the expression of genomic regions harboring HERV-W elements that can be further explored by methods of sufficient discriminatory power.

To further improve on the resolution of the analysis of melting temperatures described above, we developed a molecular beacon to control for the inherent variations in temperature in the heat-block of most thermocyclers. Used in combination with an algorithm for the identification of the normalized melting temperature, we reported an improved resolution of melting temperature analyses on the ABI Prism 7000 system (Applied Biosystems) by approximately three-fold and elimination of the systematic errors introduced by the instrument (Nellaker et al. 2007). We are currently applying this method along with element-specific assays (Yao et al. 2006) to examine the genomic origin and functional significance of the differentially expressed elements in individuals with schizophrenia.

Syncytin

One HERV-W element on chromosome 7q21 contains an intact *env* gene encoding an envelope protein that has been denoted syncytin that has a proposed role in placenta biogenesis (Blond et al. 2000; Mi et al. 2000). This proviral element, in the *ERVWE1* locus, appears to be a bona fide gene domesticated during human evolution (Mallet et al. 2004). Syncytin is expressed at the feto-maternal interface (Malassine et al. 2005), an area characterized by immunological conflict between mother and fetus. Elsewhere, expression of syncytin is very limited during normal conditions. Interestingly, aberrant expression of this particular *env* gene has previously been associated with preeclampsia (Chen et al. 2006; Keith et al. 2002; Knerr et al. 2002; Lee et al. 2001), multiple sclerosis and motor neuron disease (Oluwole et al. 2007).

Studies of postmortem brains have indicated that individuals with schizophrenia and related disorders have altered levels of the HERV-W gag protein (Weis et al. 2007a) and of the HERV-W receptor neutral amino acid transporter 1 (ASCT1; Weis et al. 2007b) within several different brain regions as compared to controls. The levels of HERV-W transcripts have also been found to be increased in some individuals with these disorders (Yolken et al. 2000) but not others (Frank et al. 2005).

The study of the control of HERV-W expression has also been evaluated in cell lines. In our experimental study on human cell lines, transcription of the *env* gene encoding syncytin was strongly induced by influenza A virus infection. Studies on the transcriptional regulation of syncytin have identified basal promoter activity in the U3 region of the 5'-LTR as well as enhancer activity in flanking genomic DNA upstream of the LTR (Prudhomme et al. 2004). In this upstream regulatory region, binding sites for several different transcription factors have been identified, including binding sites for the transcription factor, glial cells missing (Gcm) 1. Overexpression of Gcm1 in cells of placental origin has also been reported to induce fusion and elevated levels of syncytin transcripts suggesting a functional role of Gcm1 in the transcriptional regulation of syncytin. This role is also supported by the very restricted expression of Gcm1, detectable only in the human placenta and adult parathyroid.

Glial cells missing

Gcm1 is the mammalian analog of the Gcm identified in Drosophila. Gcm mutant flies were reported to lack glial cells. Subsequent studies identified Gcm as a binary switch in the developing CNS, determining glial or neuronal fate of neural precursor cells (reviewed in Jones 2005). Experimental studies in mice (Iwasaki et al. 2003) and chickens (Soustelle et al. 2007) suggest that, in higher order species, the role of Gcm in CNS development may be conserved and that vertebrate Gcm may determine glial and/or neuronal fate depending on the cellular context (Soustelle et al. 2007). Whether or not syncytin is also expressed during certain stages of human CNS development has not been studied. Dupressoir and coworkers (2005) reported on the identification of two retroviral *env* genes containing intact open reading frames in the mouse genome. These were reported to be expressed at high levels only in the placenta and were suggested to constitute functional murine analogs of human syncytin. These genes were therefore denoted syncytins A and B. Based on our findings regarding transactivation of HERV-W elements by influenza A virus, including that encoding syncytin (Nellaker et al. 2006), we investigated if one or both of the proposed mouse analogs would respond similarly. We found syncytin B, but not syncytin A, to be greatly induced by the virus infection in NIH-3T3 cells as well as in primary hippocampal neurons and glia (Asp et al 2007). If regulated by mechanisms analogous to those observed in human cells, a concomitant transcriptional up-regulation of mouse GCMa would be expected in such infected cultures. Indeed, elevated levels of transcripts encoding mouse GCMa were induced in all three cell systems following virus infection as well as *in vivo* during the acute stages of a CNS infection with a mouse-adapted neurotropic strain (WSN/33) of influenza A virus. Cloning and overexpression of GCMa did induce expression of syncytin B, but not syncytin A, in NIH-3T3 cells, suggesting that the mouse syncytin B is a functional analog to human syncytin. Our observation of readily detectable levels of transcripts encoding GCMa and syncytin A and B in fetal brains supports the notion that Gcm, and perhaps also syncytin B, may play a role during development of the mammalian brain.

Antibodies to Retroviral Proteins in Individuals with Psychiatric Disorders

The measurement of antibodies in the blood of individuals with psychiatric disorders remains an important tool for the study of the role of potential antigens in the etiopathogenesis of these disorders. In a preliminary study, we showed that individuals with recent onset schizophrenia have increased levels of antibodies to purified virions derived from several group D retroviruses, including Mason-Pfizer monkey virus (MPMV), baboon endogenous virus (BaEV) and simian retrovirus type 5 (SRV-5; Lillehoj et al. 2000). However, the specificity of the reactivity to cell culture-derived virions is difficult to document with certainty. For this reason, we adapted methods for the cloning and expression of retroviral proteins in a manner that facilitates the accurate measurement of antibodies in large numbers of clinical samples. Initial testing was performed using antigens derived from simian retrovirus type 1 based on the results of the preliminary study.

GST fusion proteins of the SRV1 *gag* and *env* were expressed in insect cells using recombinant baculovirus. The SRV1 gag and env (excluding the signal peptide) coding sequences were artificially synthesized (GenScript Corporation, Piscataway, NJ) and codon-modified for optimal expression in insect cells. The genes were designed to introduce BamH1 and EcoR1 restriction sites at the 5' and 3' ends, respectively, of the coding sequence and cloned into the corresponding restriction sites of the baculovirus transfer vector pAcSecG2T (BD Biosciences, San Jose, CA) downstream of the polyhedrin promoter and the GST open reading frame. An initial recombinant baculovirus stock was generated by co-transfection of *Spodoptera frugiperda* sf9 insect cells with the transfer vector and the ProEasy linear baculovirus DNA (AB Vector, San Diego, CA) by using Cellfectin (Invitrogen) according to the procedure suggested by the manufacturer of ProEasy. Large-scale production of recombinant proteins was performed as previously described (Viscidi et al. 2003). Briefly, approximately 5×108 *Trichoplusia ni* (High Five) cells (Invitrogen, Carlsbad, CA) were infected with 2.5 ml of a high-titer recombinant baculovirus stock in 20 ml of Ex-Cell 400 medium (JRH Biosciences, Lenexa, KS) for 60 min at room temperature with periodic inversion. Aliquots of infected cells (1×10^8) were grown as adherent cultures in tissue culture plates (245 by 245 mm; Nunc, Naperville, IL) in a volume of 100 ml of Ex-Cell 400 medium supplemented with gentamicin (10 g/ml). After 96 h of incubation at 27°C, the cells were harvested from the plates by scraping and were collected by centrifugation at 2 000 rpm (Sorvall FH18/250 rotor) for 5 min. The cells were lysed in PBS containing 0.5% NP-40 and a cocktail of protease inhibitors (Roche). After brief sonication, clarified cell lysates were obtained with high-speed centrifugation (15 000 xg). The final protein concentration of the clarified lysates was adjusted to 3 mg/ml with PBS 0.05% Tween-20 and aliquots stored at −70°C.

The GST-tagged proteins were bound to casein-coated microtiter plate as outlined in Fig 2. Preparation of glutathione-casein and GST capture ELISA was performed as previously described (Sehr et al. 2001) with minor modifications. In brief, casein (Sigma Chemical Company, St Louis) at a concentration of 5 mg/ml in (PBS) was incubated for 15 min at room temperature (RT) with 0.4 mM *N*-ethylmaleimide (NEM; Sigma). Subsequently, 4 mM sulfosuccinimidyl 4-[p-maleimidophenyl]butyrate (SSMBP; Pierce, Rockford, IL) was added as crosslinker and the reaction proceeded for 30 min at RT. Free SSMBP and NEM were separated from casein by size exclusion chromatography on PD10 columns (Pharmacia). The protein fraction was then supplemented with 10 mM glutathione (Sigma) and the coupling reaction was executed for 1 h at RT. The glutathione-casein was separated from unbound glutathione by gel filtration with PD10, using PBS as buffer, and was stored at −20°C in small aliquots.

Polysorb plastic plates, 96 wells (Nunc), were coated overnight at 4°C with 200 ng/well of glutathione casein in 50 mM carbonate buffer, pH 9.6. Thereafter, wells were washed once with PBS 0.05% Tween 20 and incubated for 1 h at 37°C with 200 µl of blocking buffer (0.2% casein, 0.5% polyvinyl alcohol, 0.5% polyvinylopyrolidone, 0.05% Tween 20 in PBS). Subsequently, the plates were incubated for 2 h at RT with the cleared lysates from the recombinant baculovirus-infected insect cells overexpressing GST-Gag and env proteins diluted in blocking buffer to 0.3 mg/ml of total protein. Human sera were diluted 1/100 in blocking buffer and incubated on the ELISA plates for 1 hr at 37°C. Bound human antibodies were detected by incubating with goat anti-human IgG antibody conjugated to HRP (1/4000 dilution in blocking buffer) for 1 h at

Enzyme Immunoassay using glutathione-casein and GST-fusion proteins

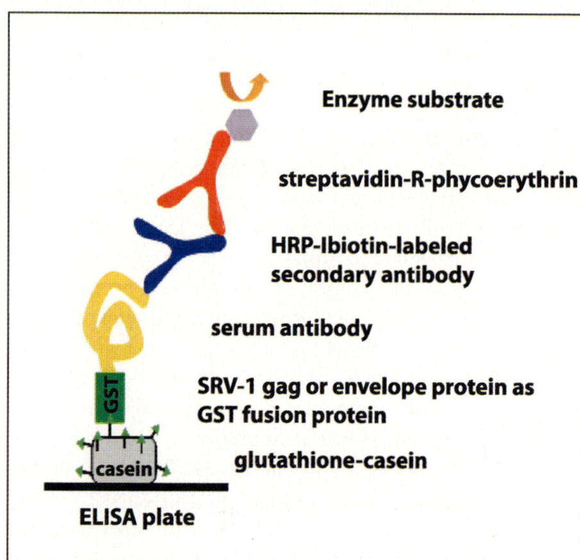

Fig. 2. Outline of method used for the binding of the GST fusion proteins to the microtiter plate and the performance of the enzyme immunoassays described in the text. (Derived from Sehr et al. 2001)

37°C. Bound reactivity was visualized by incubation with ABTS Peroxidase Substrate (KPL, Gaithesburg, MD) for 10 min at RT. The enzyme reaction was stopped by adding 100 µl of 0.5% SDS and the absorbance was measured at 405 nm. All washing steps to remove unbound reagents were performed with PBS containing 0.05% (v/v) Tween 20 in an automated plate washer. Reactivity of the human sera to lysates from baculovirus-infected insect cells expressing the GST protein alone was also measured using the same ELISA procedure described above. For each sample, an adjusted optical density was calculated by subtracting the value generated by reactivity with baculovirus proteins generated without an insert from the value generated by the baculovirus-expressed SRV-1 gag and SRV-1 envelope respectively. For each case sample, positivity was defined as yielding an adjusted optical density greater than the 90th percentile of samples from health controls run on the same mictotiter plate.

We employed the system described above to measure the levels of antibodies to baculovirus cloned simian retrovirus type 1 gag and env proteins in individuals with recent onset psychosis as well as individuals with established schizophrenia, bipolar disorder, and controls. The study population consisted of 110 individuals with recent onset psychosis, 319 individuals with established schizophrenia, 124 individuals with established bipolar disorder and 199 control individuals without a psychiatric disorder. The methods for the recruitment and evaluation of these individuals have been previously described (Dickerson et al. 2003, 2004, 2007). As depicted in Fig. 3, we found an increased level of antibodies to both envelope and gag proteins in individuals with the psychiatric disorder, particularly those with recent onset psychosis (odds ratio for individuals with antibodies to both SRV-1 gag and SRV-1 env 5.04, 95% confidence interval 1.79–14.26, p = 0.002). It is notable that the antibody levels were

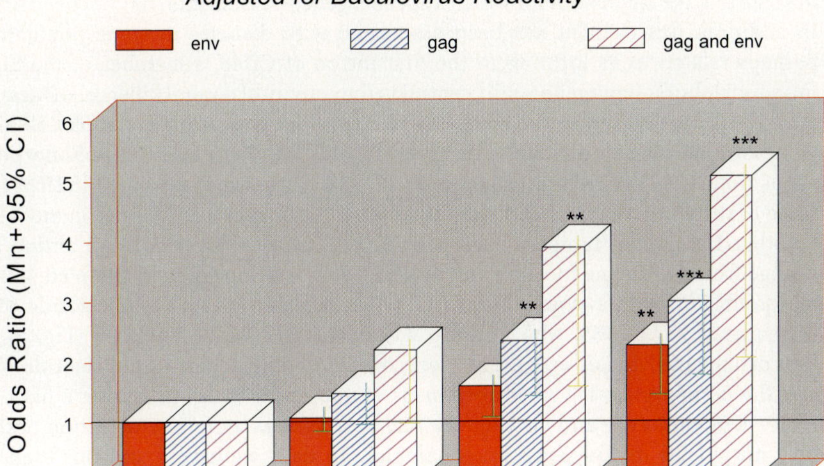

*** p<.001, **p<.02 adjusted for age, gender, race, baculovirus reactivity

Fig. 3. Reactivity to SRV-1 gag and env individuals with psychiatric disorders and controls SRV-1 gag and SRV-1 envelope proteins were cloned and expressed in baculovirus as described in the text. The adjusted odds ratios were calculated using multiple logistic regression including age, race, and gender in the regression equation

higher for individuals with recent onset psychosis than for individuals with established disorder. This finding, which is consistent with the previous antibody and PCR-based studies described above, suggests that the greatest amount of immune stimulation is occurring near the time of the onset of disease symptoms and decreases later in the course of the disease. The antigenic origin of the antibodies reactive to SRV-1 gag and SRV-1 envelope protein in the individuals with psychiatric disorders is not known with certainty. Possible sources of antigenic stimulation include HERV-W syncytin, which shares a substantial amount of homology to SRV-1 envelope protein, other endogenous human retroviruses, or exogenous group D retroviruses, which have been shown to infect humans on rare occasions (Morozov et al. 1996). The specific antigens recognized by the antibodies generated by individuals with psychiatric disorders are the subjects of ongoing studies.

Polymorphisms in HERV K18

The association between HERVs and human psychiatric disorders may also be mediated by genetic polymorphisms in HERV coding sequences. We determined the relationship between polymorphisms in the envelope gene endogenous retrovirus HERV K18 and the risk of schizophrenia. This endogenous retrovirus was selected since it is located on chromosome 1q22-23, a region that has been shown to be associated with schizophrenia in some linkage studies but for which a specific genetic association has not been found. In addition, this element has been associated with diabetes in some populations, perhaps related to its location in the first intron of CD48, a member of the SLAM immunoglobulin supergene family central to the activity of natural killer cells (Messmer et al. 2006). We first sequenced the HERV K18 envelope gene in brain samples obtained postmortem from 83 individuals (Torrey et al. 2000). We identified a G/A polymorphism at position 86204 in GenBank sequence AL121985.13 from chromosome 1 in DNA from three of the 28 brains from individuals who had schizophrenia, but not in any of the 55 other DNA brain samples. This polymorphism, which results in an amino acid change from isoleucine to valine in the HERV K18 envelope, was in complete linkage disequilibrium with a synonymous C/T polymorphism located at nucleotide 85144 of sequence AL 121985, which is also within the HERV K18 envelope (Fig. 4). We thus defined individuals with the homozygous A/A polymorphism at nucleotide 86204 and the homozygous T/T polymorphisms at nucleotide 85144 as having a high-risk HERV K18 haplotype and developed a real time PCR assay for the detection of these polymorphisms in additional brain samples. To date, we have found this high-risk haplotype in the DNA from 16 of 126 (12.7%) individuals with schizophrenia but in only 1 of 84 (1.2%) control individuals (p < 0.02 adjusted for age, race, and gender). On the other hand, this haplotype was only found in 3 of 97 (3.1%) individuals with affective disorders. This rate did not differ significantly from that of controls.

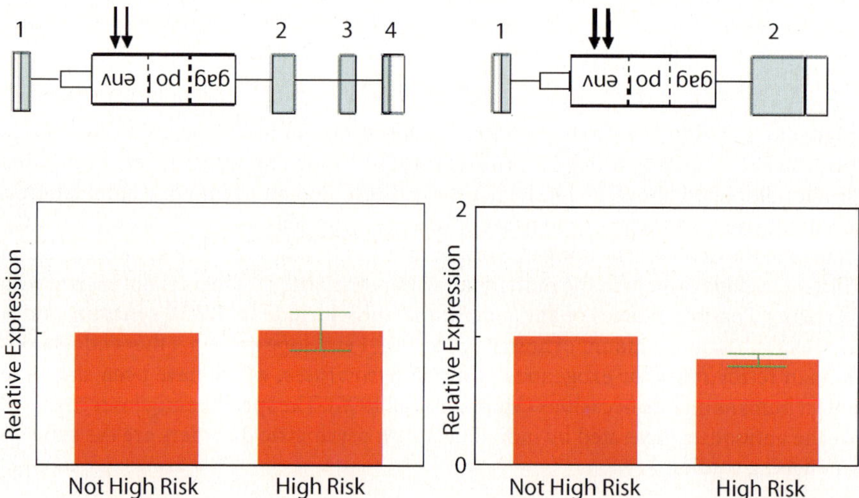

Fig. 4. Association between the HERV K18 high-risk haplotype and frontal cortex expression of two isoforms of CD48. The *bars* represent the mean and 95% confidence intervals of the expression in the frontal cortex of the 4-exon isoform of CD48 (*left*) and the 2-exon isoform (*right*) in individuals with the high-risk haplotype of HERV K18 on chromosome 1q22 (High Risk) and individuals who do not have this haplotype (Not high risk). The *arrows* indicate the location of the polymorphisms that define this haplotype within the env gene of Herv K18. The *arrows* are at nucleotides 85144 and 86204 of GenBank sequence AL 121985.13. The exon numbers of CD48 are indicated; coding regions are shown in *blue*. Note that the Herv K18 retroelement is oriented antisense to the coding of the CD84

We also examined the association between this high-risk haplotype and gene expression in the frontal cortex of 105 post-mortem brains that had been analyzed by microarray analysis (Elashoff et al. 2007). We found a strong association between the high-risk haplotype and the decreased expression of an alternate-spliced form of CD48 (expression ratio = 0.82, 95% confidence interval 0.76–0.87, p < 0.001). On the other hand, there was no association between the high-risk haplotype and expression of the fully spiced form of CD48 (Fig. 4). The high-risk haplotype was also associated with altered brain frontal cortex expression of genes in a number of different gene ontology pathways related to the immune response. Overall, of the 359 genes in the immune response gene ontology pathway on which we had data, there were 42 that showed a significantly increased association with this haplotype and one that showed a significant decrease (defined as p<0.01; Table 1). These findings are of interest in light of studies indicating a high rate of immune dysregulation in individuals with schizophrenia (Strous and Shoenfeld 2006).

Our studies indicate that a haplotype defined by polymorphisms on HERV K18 in chromosome 1q22-23 may be associated with an increased risk of schizophrenia in individuals who die with this disease. While modest, this odds ratio is higher than the odds ratios that have been found for other genetic alterations associated with schizophrenia. This risk genotype was not associated with other psychiatric diseases such as bipolar disorder, indicating a degree of specificity for schizophrenia. The mechanism by which this polymorphism increases the risk of schizophrenia is likely to be associated with the altered transcription of CD48, with subsequent changes in the immune response to antigens. The specific pathways by which these alterations lead to schizophrenia, as well as the role of these pathways in the autoimmune disorders that are associated with schizophrenia, should be the subjects of additional investigations. It is notable that the polymorphisms identified in this study and others within HERV K18 on chromosome 1 are not included in several of the panels used for "whole genome" association studies, due to their repetitive nature and the fact that they are not amenable to hybridization methods that measure small target regions. Associations between complex disorders and these regions would thus not be likely to be detected by these methods. The ability to detect genetic risks for complex disorders thus might be markedly enhanced by the development and utilization of methods capable of identifying associations with endogenous retroviruses and other repetitive regions throughout the human genome.

Conclusion

Despite intensive research efforts performed over many years in laboratories around the world, the etiopathogenesis of schizophrenia remains unclear. We postulate that detailed study of the role of endogenous retroviruses will contribute to a better understanding of this disease, as well as of other complex disorders of the human CNS. Future studies will include more detailed genetic analyses of polymorphisms in HERV regions of the human genome as well as more detailed genomic and proteomic analyses of HERV gene expression. It is notable that many of the tools that are being developed for genetic, genomic, and proteomic analyses largely exclude repetitive regions of the genome, such as those that contain HERVs. It is thus imperative that techniques be

Table 1. Genes with altered brain frontal cortex expression associated with the high-risk HERV K18 haplotype

Locus-link ID	Gene Symbol	Gene name	Ratio	Standard Error	P Value
3077	HFE	Hemochromatosis	1.15	0.03	< 0.00001
3119	HLA-DQB1	Major histocompatibility complex, class II, DQ beta 1	1.12	0.03	>0.00001
3507	IGHM	Immunoglobulin heavy constant mu	1.16	0.04	<0.00001
3107	HLA-C	Major histocompatibility, complex, class 1, C	1.29	0.06	(<0.00001
6892	TAPBP	TAP binding protein (tapasin)	1.13	0.03	0.00004
3105	HLA-A	Major histocompatibility complex, class I, A	1.17	0.05	0.00011
8741	TNFSF13	Tumor necrosis factor (ligand) superfamily, member 13	1.15	0.05	0.00013
10859	LILRB1	Leukocyte immunoglobulin-like receptor, subfamily B (with TM and ITIM)	1.06	0.02	0.00019
7940	LST1	Leukocyte specific transcript 1	1.12	0.04	0.00021
3140	MR1	Major histocompatibility complex, class I-related	1.10	0.04	0.00040
3603	IL 16	Interleukin 16 (lymphocyte chemoattractant factor)	1.15	0.05	0.00045
5871	MAP4K2	Mitogen-activated protein kinase kinase kinase kinase 2	1.09	0.02	0.00045
63940	GPSM3	G-protein signaling modulator 3 (AGS3-like, C. elegans)	1.18	0.05	0.00045
25939	SAMHD1	SAM domain and HD domain 1	1.22	0.06	0.00064
2495	FTH1	Ferritin, heavy polypeptide 1	1.15	0.05	0.00078
2204	FCAR	Fc fragment of IgA, receptor for	1.07	0.03	0.00070
355	TNFRSF6	Tumor necrosis factor receptor superfamily, member 6	1.12	0.05	0.00080
3806	KIR2DS1	Killer cell immunoglobulin-like receptor, two domains, short cytoplasmic	1.08	0.03	0.00142
4851	NOTCH1	Notch homolog 1, translocation-associated (Drosophila)	1.29	0.09	0.00170
10225	CD96	CD96 antigen	1.07	0.02	0.00200
3106	HLA-B	Major histocompatibility complex, class 1, B	1.24	0.09	0.00215
10410	IFITM3	Interferon induced transmembrane protein 3 (1-8U)	1.42	0.12	0.00218
5393	PMSCL1	Polymyositis/scleroderma autoantigen 1, 75kDa	1.15	0.06	0.00244
9567	GTPBP1	GTP binding protein 1	1.09	0.04	0.00267

Table 1. (continued)

Locus-link ID	Gene Symbol	Gene name	Ratio	Standard Error	P Value
8519	IFITM1	Interferon induced transmembrane protein 1 (9-27)	1.24	0.09	0.00280
3429	IF127	Interferon, alpha-inducible protein 27	1.17	0.06	0.00284
3113	HLA-DPA1	Major histocompatibility complex, class II, DP alpha 1	1.14	0.06	0.00299
11025	LILRB3	Leukocyte immunoglobulin-like receptor, subfamily B (with Tm and ITIM)	1.10	0.05	0.00356
5989	RFX1	Regulatory factor X, 1 (influences HLA class II expression)	1.03	0.01	0.00395
2213	FCGR2B	Fc fragment of IgG, low affinity IIb, receptor for (CD32)	1.08	0.04	0.00407
9466	IL27RA	Interleukin 27 receptor, alpha	1.12	0.05	0.00411
3903	LAIR1	Leukocyte-associated Ig-like receptor 1	1.06	0.02	0.00416
714	C1QG	Complement component 1, q subcomponent, gamma polypeptide	1.76	0.10	0.00432
4938	OAS1	2',5'-oligoadenylate synthetase 1, 40/46kDa	1.11	0.05	0.00557
970	TNFSF7	Tumor necrosis factor (ligand) superfamily, member 7	1.12	0.05	0.00624
966	CD59	CD59 antigen P18-20 (antigen identified by monoclonal antibodies 16.3A5, EJ16, EJ3, EL32 and G344)	1.09	0.04	0.00630
2533	FYB	FYN binding protein (FYB-120/130)	1.12	0.06	0.00650
80329	ULBP1	UL16ding protein 1	1.25	0.08	0.00700
972	CD74	CD74 antigen (invariant polypeptide of major histocompatibility complex)	1.17	0.07	0.00731
3570	IL6R	Interleukin 6 receptor	1.14	0.06	0.00754
3811	KIR3DL1	Killer cell immunoglobulin-like receptor, three domains, long cytoplasmic	1.08	0.04	0.00833
9437	NCR1	Natural cytotoxicity triggering receptor 1	0.92	0.04	0.00919
6387	CXCL 12	Chemokine (C-X-C motif) ligand 12 (stromal cell-derived factor 1)	1.08	0.04	0.00959

developed that will allow for the inclusion of HERVs and other repetitive sequences in the large-scale genetic analyses directed at identifying risk factors for schizophrenia. The studies presented above indicate that only with the inclusion of HERV regions will complex disorders such as schizophrenia be understood and new methods for disease diagnosis, prevention, and treatment be developed.

Acknowledgements. This work was supported by grants from the Stanley Medical Research Institute. We thank Ms Ann Cusic for her assistance in the preparation of this manuscript.

References

Abrahao AL, Focaccia R, Gattaz WF (2005) Childhood meningitis increases the risk for adult schizophrenia. World J Biol Psychiat 6(Suppl 2):44–48

Alao AO, Dewan MJ (2001) Psychiatric complications of malaria: a case report. Int J Psychiat Med 31:217–223

Anthony JM, van Marle G, Opii W, Butterfield DA, Mallet F, Yong VW, Wallace JL, Deacon RM, Warren K, Power C (2004) Human endogenous retrovirus glycoprotein-mediated induction of redox reactants causes oligodendrocyte death and demyelination. Nature Neurosci 7:1088–1095

Anthony JM, Izad M, Bar-Or A, Warren KG, Vodjgani M, Mallet F, Power C (2006) Quantitative analysis of human endogenous retrovirus-W env in neuroinflammatory diseases. AIDS Res Human Retroviruses 22:1253–1259

Asp L, Nellaker C, Karlsson H (2007) Influenza A virus transactivates the mouse envelope gene encoding syncytin B and its regulator, glial cells missing 1. J Neurovirol, 13:29–37

Atlas A, Gisslen M, Nordin C, Lindstrom L, Schwieler L (2007) Acute psychosis symptoms in HIV-1 infected patients are associated with increased levels of kynurenic acid in cerebrospinal fluid. Brain Behav Immunol 21:86–91

Bannert N, Kurth R (2004) Retroelements and the human genome: new perspectives on an old relation. Proc Natl Acad Sci USA 101:14572–14579

Behr J, Schaefer M, Littman E, Klingebiel R, Heinz A (2006) Psychiatric symptoms and cognitive dysfunction caused by Epstein-Barr virus-induced encephalitis. Eur Psychiat 21(8):521–522

Belshaw R, Katzourakis A, Paces J, Burt A, Tristem M (2005) High copy number in human endogenous retrovirus families is associated with copying mechanism in addition to reinfection. Mol Biol Evol 22:814–817

Blond JL, Beseme F, Duret L, Bouton O, Bedin F, Perron H, Mandrand B, Mallet F (1999) Molecular characterization and placental expression of HERV-W, a new human endogenous retrovirus family. J Virol 73:1175–1185

Blond JL, Lavilette D, Cheynet V, Bouton O, Oriol G, Chapel-Fernandes S, Mandrand B, Mallet F, Cosset FL (2000) An envelope glycoprotein of the human endogenous retrovirus HERV-W is expressed in the human placenta and fuses cells expressing the type D mammalian retrovirus receptor. J Virol 74:3321–3329

Bode L, Reckwald P, Severus WE, Stoyloff R, Ferszt R, Dietrich DE, Ludwig H (2001) Borna disease virus-specific circulating immune complexes, antigenemia, and free antibodies - the key marker triplet determining infection and prevailing in severe mood disorders. Mol Psychiat 6:481–491

Bohling SD, Wittwer CT, King TC, Elenitoba-Johnson KS (1999) Fluorescence melting curve analysis for the detection of the bcl-1/JH translocation in mantle cell lymphoma. Lab Invest 79:337–345

Brown AS, Susser ES (2002) In utero infection and adult schizophrenia. Ment Retard Dev Disabil Res Rev 8:51–57

Brown AS, Begg MD, Gravenstein S, Schaefer CA, Wyatt RJ, Bresnahan M, Babulas VP, Susser ES (2004) Serologic evidence of prenatal influence in the etiology of schizophrenia. Arch Gen Psychiat 61:774–780

Brown AS, Schaefer CS, Quesenberry CP Jr, Liu L, Babulas VP, Susser ES (2005) Maternal exposure to toxoplasmosis and risk of schizophrenia in adult offspring. Am J Psychiat 162(4):767–773

Buka SL, Tsuang MT, Torrey EF, Klebanoff MA, Bernstein D, Yolken RH (2001) Maternal infections and subsequent psychosis among offspring. Arch Gen Psychiat 58:1032–1037

Chen CP, Wang KG, Chen CY, Yu C, Chuang HC, Chen H (2006) Altered placental syncytin and its receptor ASCT2 expression in placental development and pre-eclampsia. Bjog 113:152–158

Chen W, Duan S, Zhou J, Sun Y, Zheng Y, Gu N Feng G, He L (2004) A case-control study provides evidence of association for a functional polymorphism – 197C/G in XBP1 to schizophrenia and suggests a sex-dependent effect. Biochem Biophys Res Commun 319:866–870

Chlabicz S, Wiercinska-Drapalo A, Dare A (1996) Clinical picture of tick-borne encephalitis among patients hospitalized in 1994 in the Department of Infectious Diseases Medical School Bialystok. Rocz Akad Med Bialymst 41:35–39

Christensen T, Petersen T, Thiel S, Brudek T, Ellermann-Eriksen S, Moller-Larsen A (2007) Gene-environment interactions in multiple sclerosis: Innate and adaptive immune responses to human endogenous retrovirus and herpesvirus antigens and the lectin complement activation pathway. J Neuroimmunol 183:175–188

Colmegna I, Garry RF (2006) Role of endogenous retroviruses in autoimmune diseases. Infect Dis Clin North Am 20:913–929

Costas J (2002) Characterization of the intragenomic spread of the human endogenous retrovirus family HERV-W. Mol Biol Evol 19:526–533

Dalman C, Allebeck P, Cullberg J, Grunewald C, Koster M (1999) Obstetric complications and the risk of schizophrenia: a longitudinal study of a national birth cohort. Arch Gen Psychiat 56:234–240

Davidson M (2002) Risk of cardiovascular disease and sudden death in schizophrenia. J Clin Psychiat 63:5–11

Delisi LE, Fleischhaker W (2007) Schizophrenia research in the era of the genome, 2007. Curr Opin Psychiat 20:109–110

Delisi LE, Smith SB, Hamovit JR, Maxwell ME, Goldin LR, Dingman CW, Gershon ES (1986) Herpes simplex virus, cytomegalovirus and Epstein-Barr virus antibody titres in sera from schizophrenia patients. Psychol Med 16:757–763

Dickerson FB, Boronow JJ, Stallings C, Origoni AE, Ruslanova I, Yolken RH (2003) Association of serum antibodies to herpes simplex virus 1 with cognitive deficts in individuals with schizophrenia. Arch Gen Psychiat 60:466–472

Dickerson FB, Boronow JJ, Stallings C, Origoni AE, Cole S, Krivogorsky B, Yolklen RH (2004) Infection with herpes simplex virus type 1 is associated with cognitive deficits in bipolar disorder. Biol Psychiat 55:588–593

Dickerson FB, Stallings C, Origoni A, Boronow JB, Sullens A, Yolken RH (2007) The association between cognitive functioning and occupational status in persons with a recent onset of Psychosis. J Nerv Ment Dis, in press

Dupressoir A, Marceau G, Vernochet C, Bénit L, Kanellopoulos C, Sapin V, Heidmann T (2005) Syncytin-A and synctin-B, two fusogenic placenta-specific murine envelope genes of retroviral origin conserved in Muridae. Proc Natl Acad Sci USA 102:725–730

Eaton WW, Byrne M, Ewald H, Mors O, Chen CY, Agerbo E, Mortensen PB (2006) Association of schizophrenia and autoimmune diseases: linkage of Danish national registers. Am J Psychiat 163:521–528

Elashoff M, Higgs BW, Yolken RH, Knable MB, Weis S, Webster MJ, Barci BM, Torrey EF (2007) Meta-Analysis of 12 Genomic Studies in Bipolar Disorder. J Mol Neurosci 31:221–244

Fallon BA, Nields JA (1994) Lyme disease: a neuropsychiatric illness. Am J Psychiat 151:1571–1583

Fellerhoff B, Laumbacher B, Mueller N, Gu S, Wank R (2007) Associations between Chlamdophila infections, schizophrenia and risk of HLA-A10. Mol Psychiat 12:264–272

Forsman A, Yun Z, Hu L, Uzhameckis D, Jern P, Blomberg J (2005) Development of broadly targeted human endogenous gammaretroviral pol-based real time PCRs Quantitation of RNA expression in human tissues. J Virol Meth 129:16–30

Frank O, Giehl M, Zheng C, Hehlmann R, Leib-Mosch C, Seifarth W (2005) Human endogenous retrovirus expression profiles in samples from brains of patients with schizophrenia and bipolar disorders. J Virol 79:10890–10891

Frank O, Jones-Brando L, Leib-Mosch C, Yolken R, Seifarth W (2006) Altered transcriptional activity of human endogenous retroviruses in neuroepithelial cells after infection with Toxoplasma gondii. J Infect Dis 194:1447–1449.

Germer S, Higuchi R (1999) Single-tube genotyping without oligonucleotide probes. Genome Res 9:72–78

Gilbody S, Lewis S Lightfood T (2007) Methylenetetrahydrofolate reductase (MTHFR) genetic polymorphisms and psychiatric disorders: a HuGE review. Am J Epidemiol 165:1–13

Graziano C, Giorgi M, Malentacchi C, Mattiuz PL, Porfirio B (2005) Sequence diversity within the HA-1 gene as detected by melting temperature assay without oligonucleotide probes. BMC Med Genet 6:36

Harasawa R, Mizusawa H, Fujii M, Yamamoto J, Mukai H, Uemori T, Asada K, Kato I (2005) Rapid detection and differentiation of the major mycoplasma contaminants in cell cultures using real-time PCR with SYBR Green I and melting curve analysis. Microbiol Immunol 49:859–863

Hovatta I, Lichtermann D, Juvonen H, Suvisaari J, Terwilliger JD, Arajarvi R, Kokko-Sahin ML, Ekelund J, Lonnqvist J, Peltonen L (1998) Linkage analysis of putative schizophrenia gene candidate regions on chromosomes 3p, 5q, 6p, 8p, 20p, and 22q in a population-based sampled Finnish family set. Mol Psychiat 3:452–457

Hsiao FC, Lin M, Tai A, Chen G, Huber BT (2006) Cutting edge: Epstein-Barr virus transactivates the HERV-K18 superantigen by docking to the human complement receptor 2 (CD21) on primary B cells. J Immunol 177:2056–2060

Huang WJ, Liu ZC, Wei W, Wang GH, Wu JC, Zhu F (2006) Human endogenous retroviral pol RNA and protein detected and identified in the blood of individuals with schizophrenia. Schizophr Res 83:193–199

Iwasaki Y, Hosoya T, Takebayashi H, Ogawa Y, Hotta Y, Ikenaka K (2003) The potential to induce glial differentiation is conserved between Drosophila and mammalian glial cells missing genes. Development 130:6027–6035

Jablensky A. (2000) Epidemiology of schizophrenia: the global burden of disease and disability. Eur Arch Psychiat Clin Neurosci 250:274–285

Jones BW (2005) transcriptional control of glial cell development in Drosophila. Dev Biol 278:265–273

Karlsson H, Bachmann S, Schroder J, McArthur J, Torrey EF, Yolken RH (2001) Retroviral RNA identified in the cerebrospinal fluids and brains of individuals with schizophrenia Proc Natl Acad Sci USA 98:4634–4639

Karlsson H, Schroder J, Bachmann S, Bottmer C, Yolken RH (2004) HERV-W-related RNA detected in plasma from individuals with recent-onset schizophrenia or schizoaffective disorder. Mol Psychiat 9:12–13

Keith JC Jr, Pijnenborg R, Van Assche FA (2002) Placental syncytin expression in normal and preeclamptic pregnancies. Am J Obstet Gynecol 187:1122–1123

Kirkbride JB, Fearon P, Morgan C, Dazzan P, Morgan K, Tarrant J, Lloyd T, Holloway J, Hutchinson G, Leff JP, Mallett RM, Harrison GL, Murray RM, Jones PB (2006) Heterogeneity in incidence rates of schizophrenia and other psychotic syndromes: findings from the 3-center AeSOP study. Arch Gen Psychiat 63:250–258

Knerr I, Beinder E, Rascher W (2002) Syncytin, a novel human endogenous retroviral gene in human placenta: evidence for its dysregulation in preeclampsia and HELLP syndrome. Am J Obstet Gynecol 186:210–213

Koutsilieri E, Scheller C, Sopper S, ter Meulen V, Riederer P (2002) Psychiatric complications in human immunodeficiency virus infection. J Neurovirol 8:129–133

Lavillette D, Marin M, Ruggieri A, Mallet F, Cosset FL, Kabat D (2002) The envelope glycoprotein of human endogenous retrovirus type W uses a divergent family of amino acid transporters/cell surface receptors. J Virol 76:642–6452

Lee X, Keith JC Jr, Stumm N, Moutsatsos I, McCoy JM, Crum CP, Genest D, Chin D, Ehrenfels C, Pijnenborg R, van Assche FA, Mi S (2001) Downregulation of placental syncytin expression and abnormal protein localization in pre-eclampsia. Placenta 22:808–812

Leweke FM, Gerth CW, Koethe D, Klosterkötter J, Ruslanova I, Krivogorsky B, Torrey EF, Yolken RH (2004) Antibodies to infectious agents in individuals with recent onset schizophrenia. Eur Arch Psychiatry Clin Neurosci 254(1):4–8

Lewis DA, Levitt P (2002) Schizophrenia as a disorder of neurodevelopment. Annu Ev Neurosci 25:409–432

Li T, Underhill J, Liu XH, Sham PC, Donaldson P, Murray RM, Wright P, Collier DA (2001) Transmission disequilibrium analysis of HLV class II DRB1, DQA1, DQB1 and DPB1 polymorphisms in schizophrenia using family trios from a Hans Chinese population. Schizophr Res 49:73–78

Lillehoj EP, Ford GM, Bachmann S, Schroder J, Torrey EF, Yolken RH (2000) Serum antibodies reactive with non-human primate retroviruses identified in acute onset schizophrenia. J Neurovirol 6:492–497

Maier W, Hofgen B, Zobel A, Rietschel M (2005) Genetic models of schizophrenia and bipolar disorder: overlapping inheritance or discrete genotypes? Eur Arch Psychiat Clin Neurosci 255:159–166

Maki P, Veijola J, Jones PB, Murray Gk, Kopenen H, Tienari P, Miettunen J, Tanskanen P, Wahlberg KE, Koskinen J, Lauronen E, Isohanni M (2005) Predictors of schizophrenia - a review. Br Med Bull 73–74:1–15

Malassine A, Handschuh K, Tsatsaris V, Gerbaud P, Cheynut V, Oriol G, Mallet F, Evain-Brion D (2005) Expression of HERV-W Env glycoprotein (syncytin) in the extravillous trophoblast of first trimester human placenta. Placenta 26:556–562

Mallet F, Bouton O, Prudhomme S, Cheynet V, Oriol G, Bonnaud B, Lucotte G, Duret L, Mandrand B (2004) The endogenous retroviral locus ERVWE1 is a bona fide gene involved in hominoid placental physiology. Proc Natl Acad Sci USA 101:1731–1736

Mameli G, Astone V, Arru G, Marconi S, Lovato L, Serra C, Sotgiu S, Bonetti B, Dolei A (2007) Brains and peripheral blood mononuclear cells of multiple sclerosis (MS) patients hyperexpress MS-associated retrovirus/HERV-W endogenous retrovirus, but not human herpesvirus 6. J Gen Virol 88(Pt.1):264–274

Marin M, Lavillette D, Kelly SM, Kabat D (2003) N-linked glycosylation and sequence changes in a critical negative control region of the ASCT1 and ASCT2 neutral amino acid transporters determine their retroviral receptor functions. J Virol 77:2936–2945

Messmer B, Eissmann P, Start S, Watzl C (2006) CD48 stimulation by 2B4 (CD244)-expressing targets activates human NK cells. J Immunol 176:4646–4650

Meyer J, Koro CE, L'Italien GJ (2005) The metabolic syndrome and schizophrenia: a review. Int Rev Psychiat 17:173–180

Meza NW, Rossi NE, Galeazzi TN, Sanchez NM, Colmenares FI, Medina OD, Uzcategui NL, Alfonzo N, Arango C, Urdaneta H (2005) Cysticercosis in chronic psychiatric inpatients from a Venezuelan community. Am J Trop Med Hyg 73:504–509

Mi S, Lee X, Li X, Veldman GM, Finnerty H, Racie L, LaVallie E, Tang XY, Edouard P, Howes S, Keith JC Jr, McCoy JM (2000) Syncytin is a captive retroviral envelope protein involved in human placental morphogenesis. Nature 403:785–789

Moises HW, Ruger R, Reynolds GP, Fleckenstein B (1988) Human cytomegalovirus DNA in the temporal cortex of a schizophrenic patient. Eur Arch Psychiat Neurol Sci 238:110–113

Moller HJ (2003) Bipolar disorder and schizophrenia: distinct illnesses or a continuum? J Clin Psychiat 64:23–27

Morozov VA, Lagaye S, Lyakh L, ter Meulen J (1996) Type D retrovirus markers in healthy Africans from Guinea. Res Virol 147:341–351

Mortensen PB, Pederson CB, Westergaard T, Wohlfahrt J, Ewald H, Mors O, Andersen PK, Melbye M (1999) Effects of family history and place and season of birth on the risk of schizophrenia. N Engl J Med 340:603–608

Mortensen PB, Norgaard-Pedersen B, Waltoft BL, Sorensen TL, Hougaard D, Torrey EF, Yolken RH (2006) Toxoplasma gondii as a risk factor for early-onset schizophrenia: analysis of filter paper blood samples obtained at birth. Biol Psychiat 61:688–693

Mutsuddi M, Morris DW, Waggoner SG, Daly MJ, Scolnick EM, Sklar P (2006) Analysis of high-resolution HapMap of DTNBP1 (Dysbindin) suggests no consistency between reported common variant associations and schizophrenia. Am J Human Genet 79:903–909

Nellaker C, Yao Y, Jones-Brando L, Mallet F, Yolken RH, Karlsson H (2006) Transactivation of elements in the human endogenous retrovirus W family by viral infection. Retrovirology 3:44

Nellaker C, Wallgren U, Karlsson H (2007) Molecular beacon-based temperature control and automated analyses for improved resolution of melting temperature analysis using SYBR I green chemistry. Clin Chem 53:98–103

Nelson PN, Lever AM, Smith S, Pitman R, Murray P, Perera SA, Westwood OM, Hay FC, Ejtehadi HD, Booth JC (1999) Molecular investigations implicate human endogenous retroviruses as mediators of anti-retroviral antibodies in autoimmune rheumatic disease. Immunol Invest 28:277–289

Oluwole OSA, Yao Y, Conradi S, Kristensson K, Karlsson H (2007) Elevated levels of transcripts encoding a human retroviral envelope protein (syncytin) in muscles from patients with motor neuron disease. ALS, in press

Pavlicek A, Paces J, Elleder D, Hejnar J (2002) Processed pseudogenes of human endogenous retroviruses generated by LINEs: their integration, stability, and distribution. Genome Res 12:391–399

Perron H, Garson JA, Bedin F, Beseme F, Paranhos-Baccala G, Komurian-Pradel F, Mallet F, Tuke PW, Voisset C, Blond JL, Lalande B, Seigneurin JM, Mandrand B (1997) Molecular identification of a novel retrovirus repeatedly isolated from patients with multiple sclerosis. The Collaborative Research Group on Multiple Sclerosis. Proc Natl Acad Sci USA 94:7583–7588

Pham HM, Konnai S, Usui T, Chang KS, Murata S, Mase M, Ohashi K, Onuma M (2005) Rapid detection and differentiation of Newcastle disease virus by real-time PCR with melting-curve analysis. Arch Virol 150:2429–2438

Pichon JP, Bonnaud B, Cleuziat P, Mallet F (2006) Multiplex degenerate PCR coupled with an oligo sorbent array for human endogenous retrovirus expression profiling. Nucl Acids Res 34:e46

Prasad KM, Shirts BH, Yolken RH, Keshavan MS, Nimgaonkar VL (2007) Brain morphological changes associated with exposure to HSV1 in first-episode schizophrenia. Mol Psychiat 12:105–113

Rantakallio P, Jones P, Moring J, Von Wendt L (1997) Association between central nervous system infections during childhood and adult onset schizophrenia and other psychoses: a 28-year follow-up. Int J Epidemiol 26:837–843

Rossler W, Salize HJ, van Os J, Riecher-Rossler A (2005) Size of burden of schizophrenia and psychotic disorders. Eur Neuropsychopharmacol 15:399–409

Rote NS, Chakrbarti S, Stetzer BP (2004) The role of human endogenous retroviruses in trophoblast differentiation and placental development. Placenta 25:673–683

Schwab SG, Hoefgen B, Hanses C, Hassenbach MB, Albus M, Lerer B, Trixler M, Maier W, Wildenauer DB (2005) Further evidence for association of variants in the AKT1 gene with schizophrenia in a sample of European sib-pair families. Biol Psychiat 58:446–450

Sehr P, Zumbach K, Pawlita MR (2001) A generic capture ELISA for recombinant proteins fused to glutathione S-transferase: validation for HPV serology. J Immunol Meth 253:153–162

Seifarth W, Frank O, Zeilfelder U, Spiess B, Greenwood AD, Hehlmann R, Leib-Mosch C (2005) Comprehensive analysis of human endogenous retrovirus transcriptional activity in human tissues with a retrovirus-specific microarray. J Virol 79:341–352

Singh AK, Singh M (2000) Ascaris psychosis: an unusual presentation of round worm infestation. J Assoc Physicians India 48:456

Soustelle L, Trousse F, Jacques C, Ceron J, Cochard P, Soula C, Giangranda A (2007) Neurogenic role of Gcm transcription factors is conserved in chicken spinal cord. Development 134:625–634

Stevens JR (2002) Schizophrenia: reproductive hormones and the brain. Am J Psychiat 159:713–719

Strous RD, Shoenfeld Y (2006) Schizophrenia, autoimmunity and immune system dysregulation: a comprehensive model updated and revisited. J Autoimmun 27:71–80

Tilluckdharry CC, Chadee DD, Doon R, Nehall J (1996) A case of vivax malaria presenting with psychosis. West Indian Med J 45:39–40

Torrey EF (1988) Stalking the schizovirus. Schizophr Bull 14:223–229

Torrey EF, Webster M, Knable M, Johnston N, Yolken RH (2000) The Stanley Foundation Brain Collection and Neuropathology Consortium. Schizophr Res 44:151–155

Torrey EF, Bartko JJ, Lun ZR, Yolken RH (2007) Antibodies to Toxoplasma gondii in patients with schizophrenia: A meta-analysis. Schizophr Bull 33:729–736

Tuke PW, Perron H, Bedin F, Beseme F, Garson JA (1997) Development of a pan-retrovirus detection system for multiple sclerosis studies. Acta Neurol Scand 169:16–21

Viscidi RP, Rollison DE, Viscidi E, Clayman B, Rubalcaba E, Daniel R, Major EO, Shah KV (2003) Serological cross-reactivities between antibodies to simian virus 40, BK virus, and JC virus assessed by virus-like-particle-based enzyme immunoassays. Clin Diagn Lab Immunol 10(2):278–85

Waku-Kouomou D, Alla A, Blanquier B, Jeantet D, Caidi H, Rguig A, Freymuth F, Wild FT (2006) Genotyping measles virus by real-time amplification refractory mutation system PCR represents a rapid approach for measles outbreak investigations. J Clin Microbiol 44:487–494

Weber HC, Schoeman JF, Nowitz A, Becker ML (1994) Case report: psychosis associated with hepatitis B. J Med Virol 44:5–8

Webster JP, Lamberton PH, Donnelly CA, Torrey EF (2006) Parasites as causative agents of human affective disorders? The impact of anti-psychoatic, mood-stabilizer and anti-parasite medication on Toxoplasma gondii's ability to alter host behaviour. Proc Biol Sci 273:1023–1030

Weis S, Llenos IC, Sabunciyan S, Dulay JR, Isler L, Yolken R, Perron H (2007a) Reduced expression of human endogenous retrovirus (HERV)-W GAG protein in the cingulated gyrus and hippocampus in schizophrenia, bipolar disorder, and depression. J Neural Transm 114:645–655

Weis S, Llenos IC, Dulay JR, Verma N, Sabunciyan S, Yolken RH (2007b) Changes in region- and cell type-specific expression patterns of neutral amino acid transporter 1 (ASCT-1) in the anterior cingulated cortex and hippocampus in schizophrenia, bipolar disorder and major depression. J Neural Transm 114:261–271

WHO World Health Report (2001) New understanding, new hope. Geneva

Wright P, Gill M, Murray RM (1993) Schizophrenia: genetics and the maternal immune response to viral infection. Am J Med Genet 48:40–46

Yao Y, Nellaker C, Karlsson H (2006) Evaluation of minor groove binding probe and Taqman probe PCR assays: Influence of mismatches and template complexity on quantification. Mol Cell Probes 20:311–316

Yolken RH, Karlsson H, Yee F, Johnston-Wilson NL, Torrey EF (2000) Endogenous retroviruses and schizophrenia. Brain Res Brain Res Rev 31:193–199

Zimmet P (2005) Epidemiology of diabetes mellitus and associated cardiovascular risk factors: focus on human immunodeficiency virus and psychiatric disorders. Am J Med 118:3S-8S

Is Psychosis Due to Retroviral/Retrotransposon Integration Close to the Cerebral Dominance Gene?

T.J. Crow[1], *J.S. Close*[1], *H.-S. Kim*[2], and *M.T. Ross*[3]

Summary

Scope is now limited for an environmental theory of the origins of psychosis. The last serious candidate – infection by an exogenous virus – is ruled out by the observation that, when two siblings become ill, they do so at the same age and not at the same time. An evolutionary theory is required to explain the persistence of disadvantageous variation across human populations. The hypothesis has been proposed that the variation associated with psychosis arises in relation to the speciation event - the genetic change that initiated *Homo sapiens* as a species and postulated to have introduced cerebral asymmetry ("the torque") as the basis of language. There are strong grounds for believing the asymmetry gene to be in the X-Y homologous class. The leading candidate is the *ProtocadherinX/Y* gene pair generated by a duplication from Xq21.3 to Yp that occurred 4.5–6.0 million years ago. This gene has been subject to accelerated evolution with 16 amino acid changing substitutions in *PCDHY* and, significantly, five changes in *PCDHX*. Here we investigate the modified viral theory that psychosis is due to a retroviral insertion close to the cerebral dominance gene by examining retroviral sequences in relation to *PCDHX/Y* within the homologous region, and retro elements that differ between the homologous blocks, i.e., ones that might represent new insertions. Although large numbers of such elements were found, we did not identify a particular element that was likely to have caused changes in the expression of either *PCDHX* or *PCDHY*. We conclude that psychosis is probably not due to retroviral insertion close to the cerebral dominance gene and that the variance in gene expression arises from some other mechanism, for example, epigenetic modification in meiosis.

Introduction

Psychosis (schizophrenia and manic depressive illness) constitutes a spectrum of illnesses with an onset in early, middle and sometimes late adult life that occurs across populations at a relatively high prevalence – 2–3% lifetime expectation. The commonly held view that these disorders are multifactorial in origin and that a number of etiological agents, each of small effect, have been identified is little more than a confession

[1] SANE POWIC, Warneford Hospital, Oxford, OX3 7JX, UK, e-mail: tim.crow@psych.ox.ac.uk
[2] Section of Biological Systems, College of Natural Science, Pusan National University, San 30, Changjeon Dong, Pusan 609-735, South Korea
[3] X Chromosome Group, Wellcome Trust Sanger Institute, Wellcome Trust Genome Campus, Hinxton, Cambridge CB10 1SA, UK

Gage et al.
Retrotransposition, Diversity and the Brain
© Springer-Verlag Berlin Heidelberg 2008

of ignorance. A genetic component is undoubted but remains ill-defined in Mendelian terms. First-degree relatives are at approximately 10% risk of illness compared to 1% for each of the main syndromes, e.g., schizophrenia or manic depressive illness, narrowly defined. But a simple genetic theory cannot account for 1) the shortfall from 100% in concordance in monozygotic twin-pairs, 2) onset in adult life, and 3) continued high prevalence against the selective effects of a reduction in fertility. To explain these facts, a gene-environment interaction is often invoked, but there is a dearth of environmental agents to which an etiological role can plausibly be attributed. Psychogenic trauma has fallen by the wayside from the failure of its advocates to specify the nature of the trauma or to construct a plausible account of its psychodynamic effects. Other categories of environmental agents can be reduced to toxins, infection and disturbances of immunity. Of these, a toxic factor appears unlikely in view of the widespread geographical distribution. Immunity has been invoked but never substantially documented. Thus the category of infection remains a salient possibility.

The concept that schizophrenia might be horizontally transmitted has a long history (Baillarger 1857; Hofbauer 1864; Wollenberg 1889), although only more recently in a microbial context. It has been argued (Crow 1983) that some findings in family studies are compatible with the notion that, in addition to a genetic predisposition, what matters is proximity to an individual who already has the disease. Consistent with this theory are the following findings:

1. Concordance rates are higher in dizygotic twins than in their siblings (Fischer 1973), although these relatives share genes to the same extent: because they are the same age. twins may be in closer contact than siblings.
2. Concordance in relatives is greater in same-sex than opposite-sex pairs (Rosenthal 1962), but this effect occurs only within the family (i.e., first-degree relatives – Penrose 1942); again it may be argued that same-sex pairs of relatives are in closer physical proximity.
3. In monozygotic twins, the second member of the pair was reported at increased risk after the onset of the illness in the first, and the increase was confined to pairs who were together at this time (Abe 1969).

These arguments were placed in the context of other evidence consistent with a viral etiology:

1. Coincidence of psychosis with infectious disease, particularly influenza

Menninger (1926) reported a series of 175 cases of post-influenzal psychosis and noted that the initial clinical picture had the "unmistakable schizophrenic stigmas including intrapsychic ataxia, emotional ideational spitting, incoherence, stereotypies and other bizarre expressions" and that these symptoms were not conspicuously different from those seen in the usual types of schizophrenic illness. He concluded that the schizophrenic syndrome i) was the relatively most frequent post-influenzal psychosis, ii) occurred with or without predisposition or hereditary taint, and iii) in most cases terminated in complete recovery. But he modified his view and later wrote, "I think I said seven years ago that influenza caused psychosis. I have changed my mind" (Menninger 1928). Others pursued the viral hypothesis with greater tenacity. Goodall

(1932), in the wake of the encephalitis lethargica epidemic that followed and may well have been related (Ravenholt and Foege 1982) to the 1918 influenza epidemic noted that, "there are observers who consider that there is no essential difference between psychotic disturbance connected with encephalitis (post-encephalitic) and those met with in states covered by the description schizophrenia" and went on to argue that, "epidemic encephalitis, with the somatic disorders which accompany it, may be a virus disease and similarly caused perhaps are the schizophrenic states which resemble them." Jelliffe (1927), Hendrick (1928), and McCowan and Cook (1928) also related some cases to the 1918 epidemic, which was almost certainly due to influenza A (H1N1), but the association appears not to have occurred or to have been documented subsequently.

2. Epidemiology

Seasonality:

Consistent with a viral etiology are seasonal changes. Data have been presented that there is an excess of onsets of both mania and schizophrenia in the early summer months (Hare and Walter 1978), expressed in relation to admissions for other types of illness, and that there is a season-of-birth effect: individuals who later develop the disease are more likely (by 4–8%) to have been born in the months of winter and early spring than at other times of the year (Torrey and Petersen 1977). Similar effects are seen for mania (Hare and Walter 1978) as for schizophrenia.

Opposing these assumptions are the arguments of Lewis (1989) that the season-of-birth effect is due to "age incidence:" the fact that those who are born in the early months of the year are at greater risk (and more exposure) than those born in the later months when the data on subsequent incidence are collected by calendar year. This argument has not been effectively eliminated, and the prediction that the season-of-birth effect would reverse in the southern hemisphere if due to a climatic factor is unclear.

Temporal Geographical Variations:

A challenge has been issued against the common assumption that the incidence of schizophrenia is uniform with respect to time, and approximately similar in the various populations of the world. Hare (1983) presented evidence on the basis of the increase in hospital facilities that incidence of the disease increased in the nineteenth century. Both Hare (1983) and Torrey (1980) have drawn attention to the relative scarcity before 1800 of case descriptions that, with reasonable confidence, can be classified as schizophrenia.

These arguments were coupled with evidence for significant variations in incidence across populations (Torrey 1980; McGrath 2006). The following considerations are relevant to claims that incidence varies over time and space:

1. In a review of the classical literature, Roccatagliata (1991) wrote that "we are forced to recognise a substantial identity between the ancient and the modern approach. There seemed to be no remarkable differences between the nosology of schizophrenic disturbances proposed by the APA Diagnostic and Statistical

Manual of Mental Disorders on the one hand and classical nosology on the other ... From a scientific viewpoint this confirms the existence above all of interpretive models and natural psychiatric entities that have been modified only marginally by historical and social development." Such findings are paralleled by relative stability of the clinical picture over time when records are available from recent centuries (Turner 1992; Nixon and Doody 2005).

2. When systematically sought for, it appears that the patterns of illness described as schizophrenia or manic depressive psychosis are relatively constant across populations. Thus, for example, Murphy (1976), who examined psychological abnormality in the Eskimo, the Yoruba in Nigeria and populations of Sweden and Canada reached the conclusion that "rather than being simply violations of the social norms of particular groups, as a labelling theory suggests, symptoms of mental illness are manifestations of a type of affliction shared by virtually all mankind." The conclusion is borne out by studies in relatively isolated and primitive populations. Thus in an investigation of the prevalence of schizophrenia in Botswana, Ben-Tovim and Cushnie (1986) concluded that prevalence was "within the range generally reported for industrial communities. Remote village life in Botswana appears to offer no protection against the development of schizophrenia." Similarly in a comparison of colored and black individuals of Bantu, Negroid and Khoisan origin with different language and cultural backgrounds, Maslowski et al. (1998) concluded that "core symptoms remained basically the same" across groups, but the content of positive symptoms varied somewhat with culture and "no statistically significant differences in the presentation of negative symptoms were found."

3. In those cases where apparently unusual rates of schizophrenia have been observed, eg Ireland, subsequent investigation has cast doubt on the exceptionality. In an investigation of the incidence and prevalence in Ireland, Cabot (1990) concluded that "cultural and genetic hypotheses have been advanced to explain figures without a critical examination of the studies at the basis of this claim ... and hypotheses are reviewed to show that their conclusions are equivocal and ungeneralisable ... most often due to the method of case collection and inconclusive across cultural comparison. It is hoped that Ireland will not henceforth be considered a high prevalence area for schizophrenia without more reliable research". Similarly Ni Nuallain (1990) concluded that the apparent higher rates for Ireland were due to "continued hospitalisation of symptomatically recovered cases that has given rise to the mistaken impression that the prevalence of schizophrenia is unduly high in Ireland ... The work reported here indicates substantial differences between results of case ascertainment by hospital admission compared with those arrived at by standardised interview diagnostic techniques". Thus considerable caution is indicated in reaching the conclusions suggested by McGrath (2005) that there are substantial and unaccounted for variations in incidence of psychosis across populations.

3. Direct Investigations of the Viral Hypothesis

Three sources of evidence are advanced as consistent with a viral aetiology:

1. Serum and antibody titers: Elevated CSF serum ratios of antibodies to cytomegalovirus (Albrecht et al. 1980; Torrey et al. 1982) and two other herpes viruses have been reported, but the most systematic studies (e.g., King et al. 1985) do not show obvious excesses.
2. Cell culture techniques: In a search for known or unrecognized viruses, cerebrospinal fluid (CSF) from about one third of patients with schizophrenia was found to induce a cytopathic effect in human embryonic fibroblast cultures (Tyrrell et al. 1979). The effect was not passaged but was prevented by filters of 50 nm pore size; it was thought that the effect might reflect the presence of a small RNA virus. Further investigation showed that the effect was also seen with CSF from some patients with affective disorder and miscellaneous neurological conditions (Taylor et al. 1985) and that it is not prevented by protein synthesis or nucleic acid synthesis inhibition. Although these findings were initially reported as consistent with the presence of a virus, the later findings made this unlikely, although they might reflect release into the CSF of a toxic factor associated with neural damage.
3. Transmission experiments: CSF inducing cytopathic effects from patients with schizophrenia was injected intracerebrally into mice and hamsters without significant effects, but when marmosets were injected, behavioral changes were observed over the next $2\frac{1}{2}$ years: the injected animals appeared less active. These experiments were subsequently repeated without the same findings being obtained (Baker et al. 1989).

A Continuum of Psychosis

One of the problems for those who maintain that there are significant variations in the incidence of either schizophrenia or manic depressive illness over time is that the boundaries are unclear. Schizo-affective disorders are common and genuinely intermediate between the more typical schizophrenic and manic depressive syndromes.

The WHO Ten Country Study of incidence and manifestations of schizophrenia (Jablensky et al. 1992) reached the conclusion that "schizophrenic illnesses are ubiquitous, appear with similar incidence in different cultures, and have clinical features that are more remarkable by their similarity across cultures than by their differences." Similar conclusions have been drawn concerning affective illness by Weissman et al. (1996). Parallel to these are the findings of population surveys (van-Os et al. 1997). Symptoms comparable to those seen in the typical syndromes are present in relatively high frequency in the general population and have the same correlates. Thus it is possible that we are dealing with continua of variation rather than discrete disease states.

Following extension of the contagion hypothesis to the case of siblings who both develop psychosis, a test of horizontal transmission was devised. If transmission occurs from one predisposed (e.g., genetically) sibling to another or if both encounter an exogenous agent at the same time, illness onset will be expected to be earlier in the younger than in the older sibling, i.e., onsets will occur at the same time and not at the same age. While the initial observation suggested that this might be the case (Crow and Done 1986), it was soon appreciated that an artefact enters the collection of such data as follows: if information is collected at the time of onset of illness in one sibling, then one is likely to hear about younger siblings who have already become ill but not those who become ill later. The artefact can be removed by exclusion of the data to elder sibling ill first pairs. When this is done the apparent bias disappears and the conclusion is clear: onsets of illness in pairs of siblings occur on average at the same age and not at the same time. This conclusion is awkward, not only for a contagion hypothesis but also for any theory that invokes an environmental agent as relevant to onset of illness.

A New Hypothesis: Retroviruses and Proto-Oncogenes

Confronted with these findings. it appeared that a radical reorientation was required. The genetic component appeared more substantial and an environmental precipitant was apparently excluded, at least in the majority of cases.

Nevertheless some attractions of the viral hypothesis remained. It had the potential to explain deleterious outcome and perhaps even the brain changes that were by this time well documented in imaging (Johnstone et al. 1978) and post-mortem (Brown et al. 1986) studies and also the sometimes relatively abrupt onset and episodicity. Affective illnesses are often unequivocally episodic in nature and the same is true to a lesser extent of those at the schizophrenic end of the spectrum. What could account for this apparent lability if the basic predisposition was genetic?

A new finding also had to be accommodated:the brain changes, including particularly ventricular enlargement, were at least in part lateralized. In a post-mortem study (Brown et al. 1986), thinning of the parahippocampal gyrus was observed in those who had suffered from schizophrenic psychoses relative to those with affective illness that was selective to the left side (side-by-diagnosis interaction, p<0.02). The hypothesis was proposed (Crow 1984) that: *"Schizophrenia (and perhaps manic depressive psychosis) is due to infection with a virus that becomes integrated in the genome sometimes to be passed from one generation to the next..*

The hypothesis took its origin in work on retroviruses (Weiss 1978), the discovery of the enzyme reverse transcriptase (Baltimore 1969; Temin and Mizutani 1970) and to models in which viral RNA was reverse transcribed into the host genome in the form of a provirus (Lwoff 1965).

In its initial formulation, the hypothesis was related particularly to familial cases and to the season-of-birth effect. However, neither of these restrictions appears necessary. The hypothesis was thought also to have to address the problem of low concordance rates (36–58%) in cases of monozygotic twins, and the necessity for genomic integration of a virus at a consistent location in the human genome to produce the consistencies (e.g., age of onset, brain changes) seen in psychosis. These difficulties were resolved by the further hypothesis that the retroviral sequence integrates at a particular locus, and this integration was predicted to be a factor associated with the growth of the brain. Particular emphasis was laid on laterality. The hypothesis thus predicted that the retrovirus had a particular affinity for the growth factor for cerebral asymmetry.

Thus the hypothesis was formulated: "*The cerebral asymmetries underlying later-ality are established by the trophic effects of a proto-oncogene, and that the retrovirus responsible for psychosis interacts with the cellular oncogene to elicit enhanced activity, which may sometimes be destructive.*"

Inherent in this formulation was the notion that psychosis was human-specific and perhaps related to the evolution of *Homo sapiens*.

In 1987 the hypothesis was modified as follows: "*(it) is proposed that psychosis results from aberrations of the genetic programme for the development of cerebral asymmetries, these asymmetries being a particular feature of human evolution contributing to the capacity for communication and social interaction. The development of cerebral asymmetries is postulated to depend upon a specific interaction between a genetic sequence which has potential autonomy (e.g., a retrovirus or other mobile genetic element) and a growth factor or proto-oncogene. The interaction ... and the persistence of the psychoses (and their variation of form) is viewed as an unfortunate by-product of this hotspot of genetic change*" (Crow 1987).

Familial Prion Disease as a Paradigm for Transmissible Neurodevelopmental Aberrations

Parallel to and interacting with this line of thought were our investigations of familial prion disease, the Gerstmann-Sträussler syndrome (Baker et al. 1990; Masters et al. 1981). Marmosets inoculated intracerebrally with brain tissue from a woman with Gerstmann-Sträussler syndrome developed an encephalopathy indistinguishable from that seen in marmosets inoculated with brain tissue from a typical case of Creutzfeldt-Jakob disease. It was also concluded that "As in Huntington's disease, in the pedigree of the patient with Gerstmann-Sträussler syndrome women who subsequently develop the illness had increased fecundity. That the pathogen in human transmissible dementia may arise from a sequence (which itself sometimes confers the selective advantage) located within the human genome" (Owen et al. 1989).

With the identification of the prion gene by Stanley Prusiner, it became possible to investigate sequence variation in relation to disease. The first mutation to cause a neuropsychiatric syndrome was an insertion in the gene that tracked with the disease within a family subsequently demonstrated to have extra repeats in the octapeptide repeat sequence (Owen et al. 1989; Poulter et al. 1992). Another family was found to have a single base pair change at amino acid 102, and this change was shown to be significantly linked with disease within two families (Hsiao et al. 1989).

Thus a disease with a Mendelian pattern of transmission and an age of onset apparently influenced by developmental factors can be shown to be due to an element integrated in the human genome that also acts as an autonomous infectious agent. Could this be a paradigm for psychosis?

The Central Paradox and the Origin of Modern *Homo sapiens*

The question was first formulated by Huxley et al. (1964) "If schizophrenia is genetic in origin why are the genes not selected out?" There is no doubt that schizophrenia is associated with a substantial fertility disadvantage, and the same is true of the affective component of the spectrum (Stevens 1969). These questions can be put alongside another question - how old is the genetic variation? – in the context of the Out of Africa hypothesis of modern *Homo sapiens* (Stringer and McKie 1996), which argues that our species arose some 100 000–150 000 years ago in Africa and subsequently expanded to occupy five continents.

What explains the success of our species and its propensity to psychosis? Concerning the latter, if one has variation that is present at this point in time and seeks its origin, that clearly cannot be before the separation of the populations of the world, i.e., before the diaspora out of Africa, and this takes us close to the origin, i.e the speciation event. The question of Huxley et al. is reopened. The answer they (Huxley et al. 1964) gave is clearly wrong, as Kuttner et al. (1967) pointed out – there is no real evidence that the balancing advantage they sought is associated with resistance to wound shock or stress as they suggested. Nor does it make sense to postulate a balancing advantage that is unrelated to the disadvantage, i.e., manifest in behavior and relating to the nervous system. Rather, as Kuttner et al. proposed, one should seek some characteristic of the species, e.g., complex social ability, intelligence or the capacity for language. These faculties are clearly related but the most specific is surely language. It is language that is most characteristic of the species and it is also this ability that can be related to aspects of neural function.

Who had an idea of the nature of the speciation event? Curiously it seems to have been Paul Broca (1877), the discoverer of language lateralization in the frontal lobes, who wrote in a festschrift for Armand de Fleury, a colleague, "Man is, of all the animals, the one whose brain in the normal state is the most asymmetrical. He is also the one who possesses the most acquired faculties. Among these faculties ... the faculty of articulate language holds pride of place. It is this that distinguishes us most clearly from the animals."

The evidence is that Paul Broca was right. Although it is controversial, two studies in particular support the anatomical thesis. In a magnetic resonance imaging study, Gilissen (2001) observed that the torque, the putative defining characteristic of the human brain (Yakovlev and Rakic 1966), is absent in the brain of the chimpanzee. Buxhoeveden et al. (2001) found that the asymmetries of the minicolumn structure of the planum temporale, demonstrated to be asymmetrical to the left in a majority of individuals by Geschwind and Levitsky (1968), are not present in the brains of the chimpanzee and rhesus monkey. Parallel to this finding is evidence that directional handedness on a population basis, a possible correlate of lateralization of the brain and the capacity for language, is not present in the chimpanzee (Marchant and McGrew 1996; Provins et al. 1982).

Asymmetry as a Determinant of Human Ability

Annett (2002) has consistently maintained that cerebral asymmetry on a population basis is the characteristic that defines *Homo sapiens* and provides the basis for language and also that the genetics of asymmetry is the major determinant of human ability.

The evidence from the National Child Development Study (Crow et al. 1998) strongly supports these contentions. In this cohort, degrees of lateralization were assessed in terms of relative hand skill: the ability to tick squares with the right or the left hand in one minute. Those who were strongly lateralized were at an advantage with respect to those who were close to ambidexterity (the point of "hemispheric indecision") in verbal and non-verbal ability, reading and mathematical skills. For verbal ability, there was a substantial sex difference in favor of females, but the relationship to laterality was the same in the two sexes. The effect could not be attributed to a general disadvantage of those with poor hand skill as, when plotted as a function of left against right, verbal ability was found to be significantly impaired in those with good hand skill but lacking lateralization (Leask and Crow 2001).

These findings have recently been replicated by Peters et al. (2006) in the large (n = 255 000 subjects) internet survey conducted for the BBC. Those who described themselves as ambidextrous were at a disadvantage relative to those who were more strongly right- or left-handed for spatial and also verbal ability, with males having an advantage for the former and females for the latter. It appears that the genetics of lateralization is the major factor in the evolution of the human brain.

Schizophrenia as an Anomaly of Development of Cerebral Asymmetry

In the meantime, evidence has accumulated that the structural changes in psychosis are located in the cerebral cortex and relate to aspects of asymmetry. The form of the asymmetry – from right frontal to left occipital, described as cerebral torque – has implications for pathophysiology. Aspects of anatomical asymmetry are deviant in individuals who suffer from psychosis (see Crow 1990, 1997b; Esiri and Crow 2002), and there is evidence from a study of handedness in childhood that such individuals are lateralizing less, or more slowly (Crow et al. 1996; Leask and Crow 2005), than the population as a whole. In post-mortem studies, the anatomical changes appear to be more posterior in the brain; losses or reversals of asymmetry have been detected in fusiform, parahippocampal (McDonald et al. 2000), and superior temporal (Highley et al. 1999a) gyri. A curious feature of these findings is that, although the loss or reversal is present in both sexes, the relationship to age of onset differs: greater anomaly of asymmetry is related to earlier age of onset in females but later age of onset in males.

Other sex differences have been detected. Density of fibers in the corpus callosum was greater in females than in males, consistent with the generalization that connectivity is inversely related to degree of asymmetry (Highley et al. 1999b). In patients with psychosis, fiber density was reduced relative to female controls whereas in males it was increased relative to male controls. Consistent with these findings, in a magnetic resonance imaging (MRI) study (Highley et al. 2003) white matter in the occipito-temporo-parietal regions was greater in females than in males, whereas in female patients it was reduced and in male patients increased by 50% relative to same-sex controls.

Thus there are morphological changes in the brain in psychosis that are sex-dependent and may relate to the sex difference in age of onset: onsets are earlier in males and earlier onsets are generally associated with poorer outcome and preponderance of negative symptoms (affective flattening and poverty of speech). Conversely, with increasing age of onset, the proportion of females increases and the form of the illness is more likely to be paranoid, i.e., delusional. An obvious hypothesis of the sex difference in age of onset is that it is developmental, but this argument encounters the difficulty that the difference is in the wrong direction: brain size (Kretschmann et al. 1979) and verbal ability (Crow et al. 1998) develop faster in females than in males, yet onset of psychosis is earlier in males.

In frontal regions, no gross asymmetry and no change in gyral volume was detected in schizophrenia (Highley et al. 2001), but in the density of cells in the cortex (in area 9), there was an asymmetry (greater cell density) to the left in controls and loss or reversal of this asymmetry in patients (Cullen et al. 2006).

XY homology and the Xq21.3/Yp translocation

The most fundamental prediction of the asymmetry hypothesis is that the genetics of psychosis is the genetics of the speciation of *Homo sapiens* (Crow 1997a); in other words, the genetics of asymmetry is conceived as the species-defining characteristic. To approach such predictions, it is necessary that the cerebral dominance gene or right shift factor be identified.

A clue comes from sex chromosome aneuploidies. Individuals who lack an X chromosome (XO, Turner's syndrome) have non-dominant hemisphere (spatial) deficits on cognitive testing. Individuals with an extra X (XXY, Klinefelter's and XXX syndromes) have verbal or dominant hemisphere deficits (Table 1). A possible explanation is that an asymmetry determinant is present on the X chromosome. But then the question arises of why males who only have one X chromosome do not have spatial deficits such as are seen in Turner's syndrome. The answer must be that the copy of the gene on the X chromosome is complemented by a copy on the Y, i.e., that the gene is in the X/Y homologous class and that the copy on the X is protected from inactivation on the inactive X (Crow 1993). A hormonal explanation will not account for the similarity of the changes in XXY individuals, who are male, and XXX individuals, who are female. The case that the gene is present also on the Y chromosome is strongly reinforced by the verbal deficits/delays that are observed in XYY individuals (Geerts et al. 2003).

The hypothesis is further strengthened by evidence that Turner's and Klinefelter's syndrome individuals have corresponding deviations in anatomical asymmetry (Rezaie et al. 2004) and by the demonstration of a same-sex concordance effect: the tendency for handedness and sex to be associated above chance expectation, which is the hallmark of X-Y linkage (Corballis et al. 1996). A role for an X-Y homologous gene is consistent with the presence of a sex difference: brain growth is faster (Kretschmann et al. 1979) and lateralization is stronger (Crow et al. 1998) in females.

Table 1. Neuropsychological impairments associated with sex chromosome aneuploidies

	XX	XY	XO	XXY	XXX	XYY
	Normal female	Normal male	Turner's syndrome	Klinefelter's syndrome		
Number of sex chromosomes	2	2	1	3	3	3
Verbal ability	normal	normal	normal	delayed	delayed	delayed
Spatial ability	normal	normal	decreased	normal	normal	normal

Where might such a gene be located? A major chromosomal rearrangement took place in the course of hominin evolution. Two regions on the human Y chromosome short arm share homology with a single region on the human X chromosome long arm (Xq21.3; Page et al. 1984; Lambson et al. 1992; Sargent et al. 1996). These homologies were created by the duplication of a 3.9 Mb contiguous block of sequences from a chimpanzee hominin-precursor X chromosome to the Y chromosome short arm, and the transposed block was subsequently split by a paracentric inversion (by a recombination, presently undated, of LINE-1 elements; Schwartz et al. 1998; Skaletsky et al. 2003) to give two blocks of homology in Yp11.2, Fig. 1). Genes within this region are therefore present on both the X and Y chromosomes in *H. sapiens* but on the X alone in other great apes and primates.

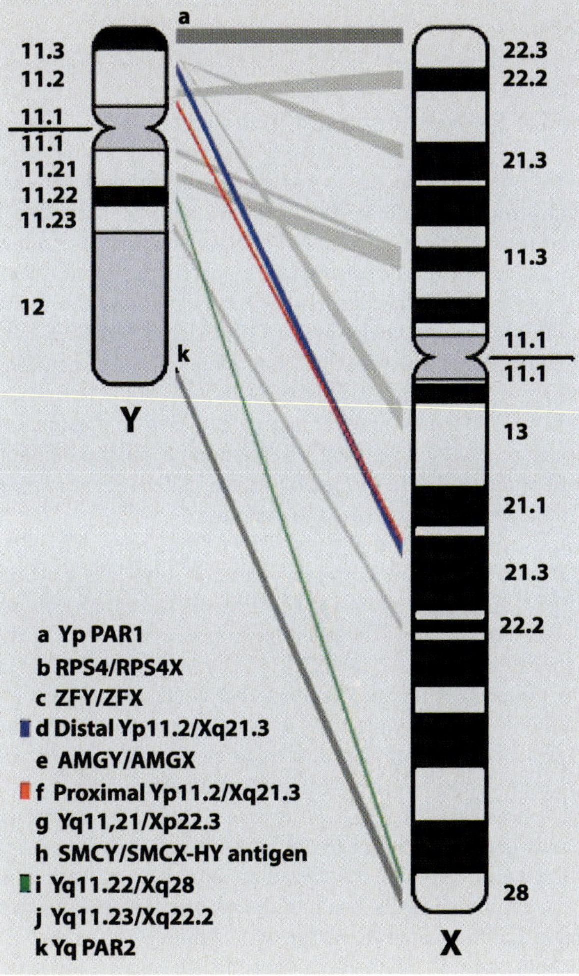

Fig. 1. Regions of homology between the X and Y chromosomes according to Affara et al (1996). The block marked in red with a sub-band in blue arose 4.5 to 6 million years ago by a duplication from Xq21.3 and Yp11.2 that, it is argued here, initiated the hominin lineage

Why were these Xq21.3/Yp11.2 homologous block sequences retained on Yp? Three genes are known to be expressed within the transposed block: *PABPC5*, a poly (A)-binding protein whose Y gametologue has been lost during hominin evolution; *TGIF2LX* and Y (homeobox-containing genes with testis-specific expression), and *ProtocadherinX (PCDH11X)* and *ProtocadherinY (PCDH11Y)*. *PCDH11X* and *Y* (each comprising seven extracellular cadherin motifs, a short transmembrane domain, and an intracellular cytoplasmic tail), which code for cell surface adhesion molecules of the cadherin superfamily, are of note because both forms of the gene have been retained and are highly expressed in fetal and adult brain (Yoshida and Sugano 1999; Blanco et al. 2000) including the germinal layer of the cortex (T.H. Priddle, personal communication). The protein products of this gene pair are thus expected to play a role in intercellular communication, perhaps acting as axonal guidance factors and influencing the connectivity of the cerebral cortex. These genes may thus have been subject to selective pressure relating to one or more brain characteristics during hominin evolution (Williams et al. 2006).

Xq21.3/Yp as a Test Bed for Retrotransposon Insertion

As a result of the duplicative translocation, we have a block of the X chromosome that is ancient in mammalian evolution and a duplicate block on the Y chromosome. Genes and sequences within the latter thus evolve free from the selective pressures that apply to the former. Thus genetic changes in the two regions acquire particular significance, first with respect to the pressures leading to the maintenance of the block on the Y, and second as an index of rates of change of sequences on the X and Y chromosomes.

There have been major rearrangements within the Y homologous block, including 4 deletions and a paracentric inversion (Fig. 2). One deletion (see Fig. 2) removed the sequence of *PABPC5* from the Y chromosome. This gene is therefore presumably irrelevant to the selective pressures maintaining the Y sequences. A second deletion removed exons 7 and 8 (both short and free of motifs) in the *PCDH11Y* sequence. However, the gene remains in frame and is expressed in the brain.

The *TGIFLX/Y* gene pair is expressed in the testis. *TGIFLX* has been subject to positive selection but this did not change in hominin evolution, and it is possible that *TGIFLY* is inactivated by a frame shift mutation. This leaves the *Protocadherin PCDH11X/Y* gene pair as the salient target of the selective pressure maintaining the block on the Y. The fact that there are changes in the PCDH11X as well as the PCDH11Y sequence (see page 99) can be adduced as further evidence that PCDH11Y is active.

The structure of *Protocadherin 11X* is represented in Fig. 3. Like other *Protocadherins* it has ectodomain repeats, in this case seven, a transmembrane domain and a cytoplasmic domain that includes a potential beta-catenin binding site and a binding site for protein phosphatase 1A, as well as a dodecapeptide repeat of obscure functional specificity that is unique to this gene. The beta-catenin binding site is of interest in view of the observations of Chen and Walsh (2003) that, if an extra beta-catenin gene is inserted in the genome to be expressed in the brain of the mouse, then gyrification of the cortex occurs. Thus it appears that some characteristics of the human cortex can be provoked in the mouse by beta-catenin. The protein phosphatase 1A gene identifies regions of the spines that make contact with incoming axons.

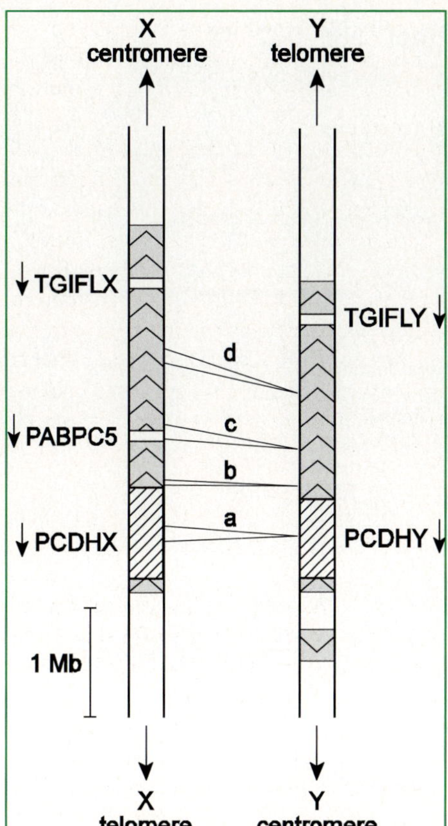

Fig. 2. Homologous blocks juxtaposed on X and Y to show the gene content with deletions marked *a–d* on the Y, deletion *a* removing exons 7 to 8 of ProtocadherinY (PCDHY) with the subsequent exons remaining in frame and expressed in fetal and adult brain. TGIFLX and Y, TG interacting factor-like gene on X and Y; PABPC5, polyadenine binding protein C5

Since the translocation 6 million years ago, there have been 15 non-synonymous, i.e., amino acid changing, substitutions on the Y chromosome. This might be expected, as the new homologous region on the Y chromosome is in a sense redundant with respect to expression of the gene on the X. But of particular interest is the fact there have also been five amino acid changing substitutions in the X sequence, which previously had been conserved in mammalian evolution. What can have accounted for this change in the hominin lineage? It seems that these changes must be a response to the presence of, and change within, the *PCDHY* sequence. In other words the *PCDHX* gene is responding to the fact that it now has a partner, *PCDHY*, with which to interact. Interestingly, three of these changes are located together within ectodomain 5 and include an arginine to cysteine substitution on the surface of the molecule. This seems likely to be reactive and therefore to signify significant structural change.

We have therefore a gene pair that is specific to hominins and has been subject to change in the hominin lineage. The question arises whether any retroviral or retro-element insertions could be relevant to the sequence of events in the evolution of *H. sapiens*. It is possible to examine this relevance because of the duplication of the Xq21.3

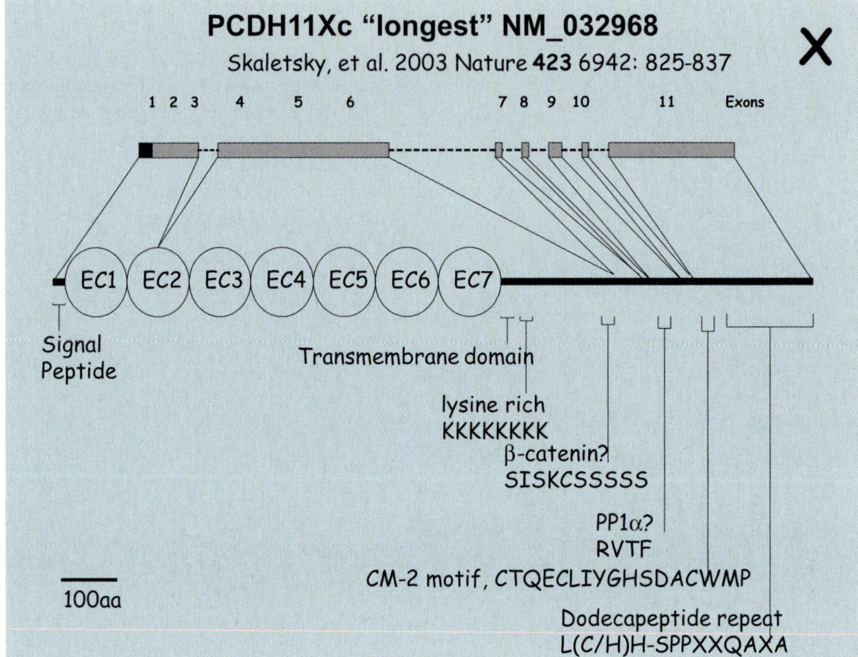

Fig. 3. Diagram representing the structure of the Protocadherin11X protein with seven extracellular repeats, a trans membrane domain, a lysine-rich region, possible beta-catenin and protein phosphatase 1a binding sites, CM-2 motif and a dodecapeptide repeat region in exon 11 of unknown function specific to this gene (From Skaletsky et al. 2003)

(using the software described by Smit, AFA, Hubley, R and Green, P. *RepeatMasker Open-3.0.1996–2004* http://www.repeatmasker.org/).

First we searched for retroviral sequences, focusing on HERV (human endogenous retrovirus) sequences, within the Xq21.3/Yp11.2 homologous region. Seven such sequences were identified (Fig. 4a): three HERV-L elements, labelled A1–A3 and four HERV-H elements, labelled H1–H4.

Six of these are conserved but the HERVHX3 sequence was removed from the Yp11.2 homologous block by a deletion that also removed exons 7 and 8 of the *PCDHY* sequence while leaving the gene in frame and expressed in the brain.

However, this event seems to be unrelated to the presence of the human endogenous retroviral sequence because the whole sequence is within the deleted block and apparently cannot have played a role in this rearrangement (Fig. 4b).

We next examined the presence of other retro-elements within the region using the RepeatMasker programme (Smit, AFA, Hubley, R and Green, P. *RepeatMasker Open-3.0.* 1996–2004 http://www.repeatmasker.org/). A number of elements are present in the X sequence that are not present in the Y, and some of these presumably represent insertions that occurred after the duplication event. The additions include a number of LINE and Alu elements, an LF SINE element and three MIR sequences (these latter are unlikely to represent de novo insertions since they became extinct with

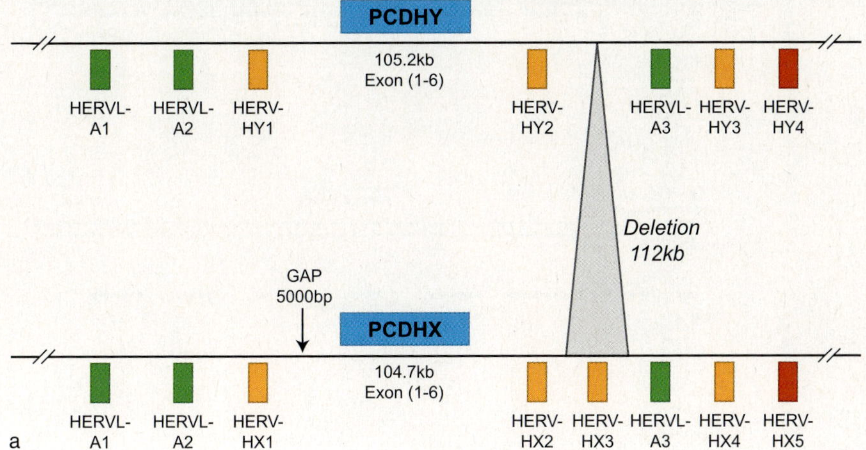

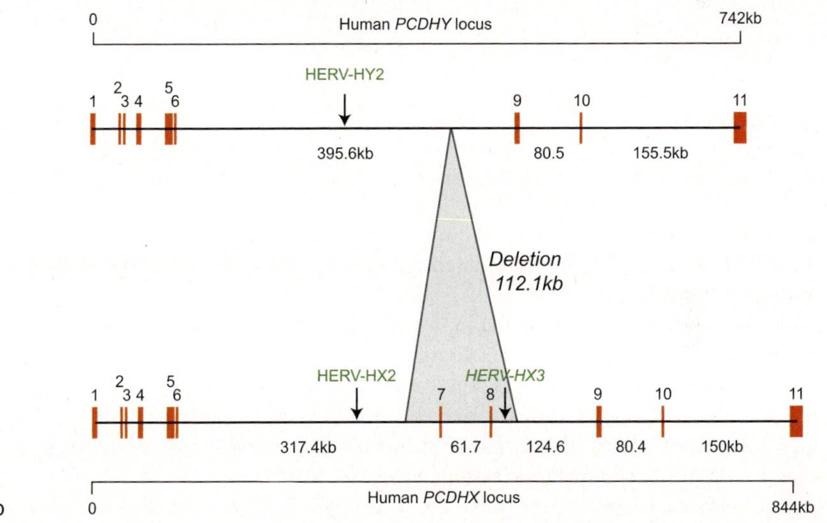

Fig. 4. (**a**) Retroviral sequences in relation to *PCDHX* and *PCDHY* gene structure within the Xq21.3/Yp11.2 region of homology. The retroviral sequences are homologous with the exception that the sequence of HERV-HX-3 on the Y was removed by the deletion that also removed exons 7 and 8 from the *PCDHY* sequence. (**b**) The location of the HERV-HX-3 sequence within the deletion and its relation to the exon structure of *PCDHY*

L2 elements). Conversely, a number of retro-elements (including L1HS sequences) are present in *PCDH11Y* that are absent from the X. Barring the possibility, yet to be investigated, that a homologous insertion has been deleted from one sequence

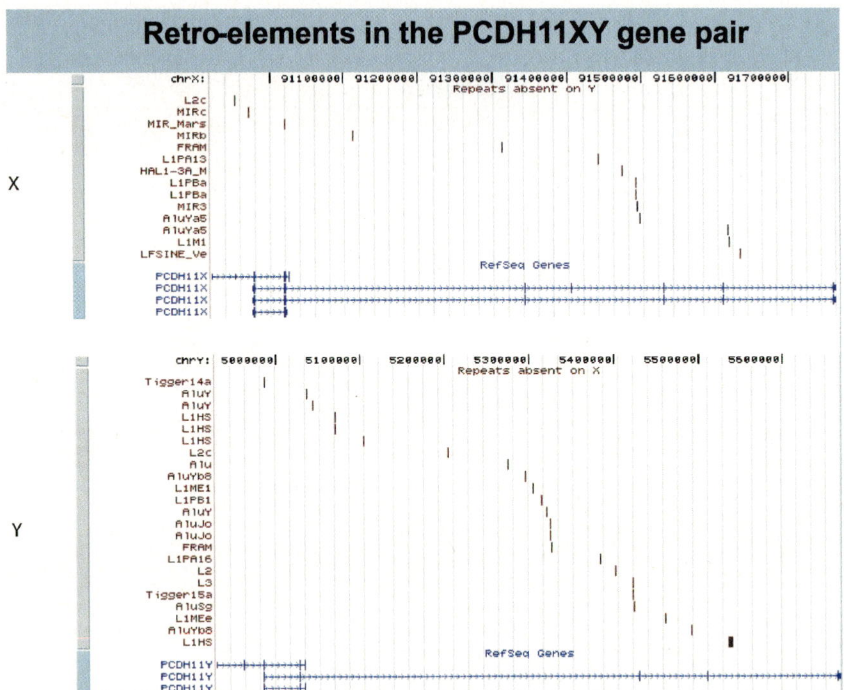

Fig. 5. Retro-elements within the *PCDH11XY* gene pair that lack homologues on the corresponding chromosome

and remains on the other, one must assume that these are new insertions in the Y sequence. They include two tigger elements, a number of LINEs L1 and Alu sequences, as represented in Fig. 5.

All these sequences are included in the region between exons 1 and 11 of *PCDHX* and *PCDHY*, respectively. While it is possible to imagine a complex scenario whereby these elements might have a role in determining alternative splice forms, it is difficult to see how such elements might affect expression of the whole protein. A simple interpretation is that these are stochastic events that happened to have occurred within this gene pair without functional consequence.

More relevant is what happens in the 5' promoter region of the gene. Here we compared the pattern of insertions in the 200 kb 5' to the first exon. Some SINE elements and DNA repeats are present within this region but these seem to be identical in the X and the Y sequences (Fig. 6). There are some repeat sequences 100 kb upstream from the start of the coding sequences and there is a large deletion from Y that is 40 kb upstream of the start, but these distances are such that the changes seem unlikely to be related to differences in gene expression. Although we have not conducted expression studies, we think it likely that there have been no significant insertions in the sequence that could have acted differentially on *PCDHX* or *PCDHY*.

In conclusion, we can probably exclude that the *PCDHX/Y* gene pair has been subject to a specific retroviral or retro-element insertion in the course of hominin

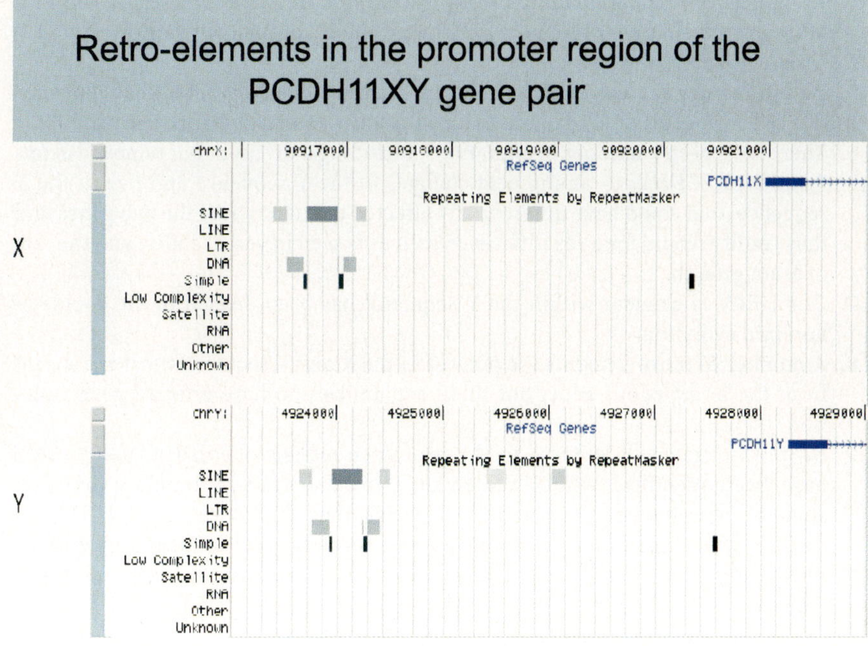

Fig. 6. Retro-elements in the promoter region 5' to the *PCDH11XY* gene pair are homologous on the X and Y, indicating an absence of evidence of novel insertions that might have influenced gene expression

evolution that selectively affected the expression of one or other of the two genes. Although one cannot exclude a more complex scenario – for example, that one or more of the elements within the gene itself or the HERV-H3 retroviral element might have some influence on pairing of the two gene sequences in male meiosis and thereby exert an influence on the expression of the gene from the inactive X chromosome – it seems that the hypothesis that an insertion in relation to the cerebral dominance gene, if the *PCDHX/Y* gene pair indeed is the correct candidate, initiated the specification of Homo Sapiens, can be excluded.

Summary

1. As Broca and Annett have supposed, cerebral asymmetry appears to be the characteristic that defines the human brain and accounts for the capacity for language.
2. Evidence from sex chromosome aneuploidies suggests strongly that the gene for asymmetry (the cerebral dominance gene) is present on both the X and the Y chromosomes.
3. Lateralization is a major determinant of verbal, non-verbal and other human abilities.

4. A duplication that occurred close to the chimpanzee hominin separation at 6 MYA created a region of homology between the X chromosome long-arm and Y chromosome short-arm.
5. Within this region, the *PCDH11X/Y* gene pair has been identified that codes for two cell surface adhesion factors that could act as axonal guidance agents.
6. There have been a number of alterations, including a deletion that removed exons 7 and 8 of *PCDHY*, 15 amino acid changes in the Y sequence and five in the X sequence, that could have differentially affected the function of the two genes and thus could account for a sex difference such as is seen in verbal ability and the rate of brain growth.
7. One HERV-H element within the Y sequence has been deleted in the course of hominin evolution.
8. A number of retro-elements are present in the X and Y sequences that are absent from the homologous gene, but these are unlikely to have affected gene function.
9. No retro-element differences in the promoter regions of *PCDHX* and *PCDHY* have been identified within the 35 kb 5' to the gene. Although expression studies have not been conducted, it seems that differences in *PCDHX* and *PCDHY* gene expression are unlikely to be secondary to de novo retroviral or retrotransposon insertion close to this candidate gene-pair for cerebral dominance.

References

Abe K (1969) The morbidity rate and environmental influence in monozygotic co-twins of schizophrenics. Br J Psychiat 115:519–531

Affara N, Bishop C, Brown W, Cooke H, Davey P, Ellis N, Graves JM, Jones M, Mitchell M, Rappold G, Tyler-Smith C, Yen P, Lau Y-FC (1996) Report of the second international workshop on Y chromosome mapping 1995. Cytogenet Cell Genet 73:33–76

Albrecht P, Torrey EF, Boone E, Hicks JT, Daniel N (1980) Raised cytomegalovirus-antibody level in cerebrospinal fluid of schizophrenic patients. Lancet ii: 769–772

Annett, M (2002) Handedness and brain asymmetry: the right shift theory. Psychology Press, Hove, Sussex

Baillarger L (1857) Example de contagion du'un delire monomanique. La Moniteur des Hospitaux 45:353–354

Baker HF, Ridley RM, Crow TJ, Tyrrell DA (1989) A re-investigation of the behavioural effects of intracerebral injection in marmosets of cytopathic cerebrospinal fluid from patients with schizophrenia or neurological disease. Psychol Med 19:325–329

Baker HF, Duchen LW, Jakobs JM, Ridley RM (1990) Spongiform encephalopathy tranmitted experimentally from Creutzfeldt-Jakob and familial Gerstmann-Straussler-Scheinker diseases. Brain 113:1891–1909

Baltimore D (1969) RNA-dependent DNA polymerase in virions of RNA tumor viruses. Nature 226:1209–1211

Ben-Tovim DI, Cushnie JM (1986) The prevalence of schizophrenia in a remote area of Botswana. Br J Psychiat 148:576–580

Blanco P, Sargent CA, Boucher C, Mitchell M, Affara N (2000) Conservation of PCDHX in mammals; expression of human X/Y genes predominantly in the brain. Mam Gen 11:906–914

Broca P (1877) Rapport sur un memoire de M. Armand de Fleury intitulé: De l'inegalité dynamique des deux hemisphères cerébraux. Bulletins de l'Academie de Medicine 6:508–539

Brown R, Colter N, Corsellis JAN, Crow TJ, Frith CD, Jagoe R, Johnstone EC, Marsh L (1986) Postmortem evidence of structural brain changes in schizophrenia. Differences in brain weight, temporal horn area, and parahippocampal gyrus compared with affective disorder. Arch Gen Psychiat 43:36–42

Buxhoeveden D, Switala AE, Litaker M, Roy E, Casanova MF (2001) Lateralization of minicolumns in human planum temporale is absent in nonhuman primate cortex. Brain Behav Evol 57:349–358

Cabot MR (1990) The incidence and prevalence of schizophrenia in the Republic of Ireland. Soc Psychiat Psychiatric Epidem 25:210–215

Chenn A, Walsh CA (2003) Increased neuronal production, enlarged forebrains and cytoarchitectural distortions in beta-catenin over-expressing transgenic mice. Cereb Cort 13:599–606

Corballis MC, Lee K, McManus IC, Crow TJ (1996) Location of the handedness gene on the X and Y chromosomes. Am J Med Genet (Neuropsychiatric Genet) 67:50–52

Crow TJ (1983) Is schizophrenia an infectious disease? Lancet 342:173–175

Crow TJ (1984) A re-evaluation of the viral hypothesis: is psychosis the result of retroviral integration at a site close to the cerebral dominance gene? Br J Psychiat 145:243–253

Crow TJ (1987) Psychosis as a continuum and the virogene concept. Br Med Bull 43:754–767

Crow TJ (1990) Temporal lobe asymmetries as the key to the etiology of schizophrenia. Schiz Bull 16:433–443

Crow (1993) Sexual selection, Machiavellian intelligence and the origins of psychosis. Lancet 342:594–598

Crow (1997a) Is schizophrenia the price that Homo sapiens pays for language? Schiz Res 28:127–141

Crow (1997b) Schizophrenia as failure of hemispheric dominance for language. Trends Neurosci 20:339–343

Crow TJ, Done DJ (1986) Age of onset of schizophrenia in siblings: a test of the contagion hypothesis. Psychiat Res 18:107–117

Crow TJ, Done DJ, Sacker A (1996) Cerebral lateralization is delayed in children who later develop schizophrenia. Schiz Res 22:181–185

Crow TJ, Crow LR, Done DJ, Leask SJ (1998) Relative hand skill predicts academic ability: global deficits at the point of hemispheric indecision. Neuropsychologia 36:1275–1282

Cullen TJ, Walker MA, Eastwood SL, Esiri MM, Harrison PJ, Crow TJ (2006) Anomalies of asymmetry of pyramidal cell density and structure in doroslateral prefrontal cortex in schizophrenia. Br J Psychiat 188:26–31

Esiri MM, Crow TJ (2002) The neuropathology of psychiatric disorder. In: Graham DI, Lantos PL (eds) Greenfield's neuropathology. London: Arnold, pp 431–470

Fischer M (1973) Genetic and environmental factors in schizophrenia. Acta Psychiatr Scand 238:1–58

Geerts M, Steyaert J, Fryns JP (2003) The XYY syndrome: a follow-up study on 38 boys. Genet Couns 14:267–279

Geschwind N, Levitsky W (1968) Human brain: left-right asymmetry in temporal speech region. Science 161:186–187

Gilissen E (2001) Structural symmetries and asymmetries in human and chimpanzee brains. In: Falk D, Gibson KR (eds) Evolutionary anatomy of the primate cerebral cortex. Cambridge: Cambridge University Press, pp 187–215

Goodall E (1932) The exciting cause of certain states, at present classified under schizophrenia by psychiatrists, may be infection. J Ment Sci 78:746–755

Hare EH (1983) Was insanity on the increase? Br J Psychiat 142:439–455

Hare EH, Walter SD (1978) Seasonal variation in admissions of psychiatric patients and its relation to seasonable variation in their birth. J Epidemiol Commun Health 32:47–52

Hendrick I (1928) Encephalitis lethergica and the interpretation of mental disease. Am J Psychiat 84:898–1014

Highley JR, McDonald B, Walker MA, Esiri MM, Crow TJ (1999a) Schizophrenia and temporal lobe asymmetry. A post mortem stereological study of tissue volume. Br J Psychiat 175:127–134

Highley JR, Esiri MM, McDonald B, Cortina-Borja M, Herron B, Cooper SJ, Crow TJ (1999b) The size and fibre composition of the corpus callosum with respect to gender and schizophrenia: A post mortem study. Brain 122:99–110

Highley JR, Walker MA, Esiri MM, McDonald B, Harrison PJ, Crow TJ (2001) Schizophrenia and the frontal lobes: A post mortem stereological study of tissue volume. Br J Psychiat 178:337–343

Highley JR, DeLisi LE, Roberts N, Webb J, Relja M, Razi K, Crow TJ (2003) Sex-dependent effects of schizophrenia: an MRI study of gyral folding, and cortical and white matter volume. Psychiat Res Neuroimag 124:11–23

Hofbauer B (1864) Infectio psychica. Österreichische Medicinische Wochenschrift (1846)39:1184–1188

Hsiao K, Baker HF, Crow TJ, Poulter M, Owen F, Terwilliger JD, Westaway D, Ott J, Prusiner SB (1989) Linkage of a prion protein missense variant to Gerstmann- Straussler syndrome. Nature 338:342–345

Huxley J, Mayr E, Osmond H, Hoffer A (1964) Schizophrenia as a genetic morphism. Nature 204:220–221

Jablensky A, Sartorius N, Ernberg G, Anker M, Korten A, Cooper JE, Day R, Bertelsen A (1992) Schizophrenia: manifestations, incidence and course in different cultures. A World Health Organization Ten Country Study. Psychol Med Suppl 20:1–97

Jelliffe SE (1927) The mental pictures in schizophrenia and in epidemic encephalitis. Am J Psychiat 6:413–465

Johnstone EC, Crow TJ, Frith CD, Steven M, Kreel L, Husband J (1978) The dementia of dementia praecox. Acta Psychiatr Scand 57:305–324

King DJ, Cooper SJ, Earle JA, Martin SJ, McFerron NV, Rima BK, Wisdom GB (1985) A survey of serum antibodies to eight common viruses in psychiatric patients. Br J Psychiat 147:137–144

Kretschmann HF, Schleicher A, Wingert F, Zilles K, Loeblich H-J (1979) Human brain growth in the 19th and 20th century. J Neurol Sci 40:169–188

Kuttner RE, Lorincz AB, Swan DA (1967) The schizophrenia gene and social evolution. Psychol Rep 20:407–412

Lambson B, Affara NA, Mitchell M, Ferguson-Smith MA (1992) Evolution of DNA sequence homologies between the sex chromosomes in primate species. Genomics 14:1032–1040

Leask SJ, Crow TJ (2001) Word acquisition reflects lateralization of hand skill. Trends Cog Sci 5:513–516

Leask SJ Crow TJ (2005) Lateralization of verbal ability in pre-psychotic children. Psych Res 136:35–42

Lewis MS (1989) Age incidence and schizophrenia: I. The season of birth controversy. Schiz Bull 15:59–73

Lwoff A (1965) Interaction among 'virus', cell and organism. Nobel Lectures in Physiology or Medicine 1963–1970. Amsterdam: Elsevier

Marchant LF, McGrew WC (1996) Laterality of limb function in wild chimpanzees of Gombe National Park: comprehensive study of spontaneous activities. J Human Evol 30:427–443

Maslowski J, Jansen van Rensburg D, Mthoko N (1998) A polydiagnostic approach to the differences in the symptoms of schizophrenia in different cultural and ethnic populations. Acta Psychiatr Scand 98:41–46

Masters CL, Gajdusek DC, Gibbs CJ (1981) Creutzfeldt-Jakob disease virus isolations from the Gerstmann-Sträussler syndrome. Brain 104:559–588

McCowan PK, Cook LC (1928) The mental aspect of chronic epidemic encephalitis. Lancet i: 1316

McDonald B, Highley JR, Walker MA, Herron B, Cooper SJ, Esiri MM, Crow TJ (2000) Anomalous asymmetry of fusiform and parahippocampal gyrus grey matter in schizophrenia: a post-mortem study. Am J Psychiat 157:40–47

McGrath J (2005) Myths and plain truths about schizophrenia epidemiology – the Nape Lecture 2004. Acta Psychiatr Scand 111:4–11

McGrath J (2006) Variations in the incidence of schizophrenia: Data versus dogma. Schiz Bull 32:195–197

Menninger KA (1926) Influenza and schizophrenia: an analysis of post-influenzal "dementia praecox" as of 1918 and five years later. Am J Psychiat 5:469–529

Menninger KA (1928) The schizophrenic syndrome as the product of acute infectious disease. Arch Neurol Psych 20:464–481

Murphy JM (1976) Psychiatric labelling in a cross-cultural perspective. Science 191:1019–1028

Ni Nuallain M, O'Hare A, Walsh D (1990) The prevalence of schizophrenia in three counties in Ireland. Acta Psychiatr Scand 82:136–140

Nixon NL, Doody GA (2005) Official psychiatric morbidity and the incidence of schizophrenia 1881–1994. Psychol Med 35:1145–1153

Owen F, Poulter M, Lofthouse R, Collinge J, Crow TJ, Risby D, Baker HF, Ridley RM, Hsiao K, Prusiner SB (1989) Insertion in prion protein gene in familial Creutzfeldt-Jakob disease. Lancet 1:51–52

Page DC, Harper ME, Love J, Botstein D (1984) Occurrence of a transposition from the X-chromosome long arm to the Y-chromosome short arm during human evolution. Nature 331:119–123

Penrose LS (1942) Auxiliary genes for determining sex as contributory causes of mental illness. J Ment Sci 88:308–316

Peters M, Reimers S, Manning JT (2006) Hand preference for writing and associations with selected demographic and behavioral variables in 255,100 subjects: The BBC internet study. Brain Cogn 62:177–189

Poulter M, Baker HF, Frith CD, Leach M, Lofthouse R, Ridley RM, Shah T, Owen F, Collinge J, Brown J, Hardy J, Mullan MJ, Harding AE, Bennett C, Doshi R, Crow TJ (1992) Inherited prion disease with 144 base pair gene insertion: I: Genealogical and molecular studies. Brain 115:675–685

Provins KA, Milner AD, Kerr P (1982) Asymmetry of manual preference and performance. Perceptual Motor Skills 54:179–194

Ravenholt RT Foege WH (1982) 1918 Influenza, encephalitis lethargica, Parkinsonism. Lancet II: 860–864

Rezaie R, Roberts N, Cutter WJ, Murphy DCM, Robertson DMW, Daly EM, Maurizio A, DeLisi LE, Crow TJ (2004) Anomalous asymmetry in Turner's and Klinefelter's syndromes - further evidence for X-Y linkage of the cerebral dominance gene. pp 102–103, 130B:102–103

Roccatagliata G (1991) Classical concepts of schizophrenia. In: Howells JG (ed) The concept of schizophrenia: historical perspectives. Washington: American Psychiatric Press

Rosenthal D (1962) Familial concordance by sex with respect to schizophrenia. Psychol Bull 59:401–421

Sargent CA, Briggs H, Chalmers IJ, Lambson B, Walker E, Affara NA (1996) The sequence organization of Yp/proximal Xq homologous regions of the human sex chromosomes is highly conserved. Genomics 32:200–209

Schwartz A, Chan DC, Brown LG, Alagappan R, Pettay D, Disteche C, McGillivray B, De la Chapelle A, Page DC (1998) Reconstructing hominid Y evolution: X-homologous block, created by X-Y transposition, was disrupted by Yp inversion through LINE-LINE recombination. Human Mol Genet 7:1–11

Skaletsky H, Kuroda-Kawaguchi T, Minx PJ, Cordum HS, Hillier L, Brown LG, Repping S, Pyntikova T, Ali J, Bierei T, Chinwalla A, Delehaunty A, Delehaunty K, Du H, Fewell G, Fulton L, Fulton R, Graves T, Hou SF, Latrielle P, Leonards S, Mardis E, Maupin R, McPherson J, Miner T, Nash W, Nguyen C, Ozersky P, Pepin K, Rock S, Rohlfing T, Scott K, Schultz B, Strong C, Tin-Wollam A, Yang SP, Waterston RH, Wilson RK, Rozen S, Page DC (2003) The male-specific region of the human Y chromosome is a mosaic of discrete sequence classes. Nature 423:825–837

Stevens BC (1969) Marriage and fertility of women suffering from schizophrenia and affective disorders. Oxford University Press, London

Stringer C, McKie R (1996) African exodus: the origins of modern humanity. J Cape, London

Taylor GR, Carter GI, Crow TJ (1985) A comparison of the effects of cytotoxic cerebrospinal fluid on cell cultures with other cytopathogenic agents. Exp Mol Pathol 42:401–410

Temin HM, Mizutani S (1970) RNA-dependent DNA polymerase in virions of Rous sarcoma virus. Nature 226:1211–1213

Torrey E F (1980) Schizophrenia and civilization. Jason Aronson, New York

Torrey EF, Petersen MR (1977) Seasonality of schizophrenic births in the United States. Arch Gen Psychiat 34:1065–1070

Torrey EF, Yolken RH, Winfrey CJ (1982) Cytomegalovirus antibody in cerebrospinal fluid of schizophrenic patients detected by enzyme immunoassay. Science 216:892–894

Turner TH (1992) Schizophrenia as a permanent problem: some aspects of historical evidence in the recency (new disease) hypothesis. History Psychiat 3:413–429

Tyrrell DAJ, Parry RP, Crow TJ, Johnstone EC, Ferrier IN (1979) Possible virus in schizophrenia and some neurological disorders. Lancet i:839–841

van-Os J, Jones P, Lewis G, Wadsworth M, Murray R (1997) Developmental precursors of affective illness in a general population birth cohort. Arch Gen Psychiat 54:625–631

Weiss RA (1978) Why cell biologists should be aware of genetically transmitted viruses. Monogr Natl Cancer Inst 48:183–189

Weissman MM, Bland RC, Canino GJ, Faravelli C, Greenwald S, Hwu HG, Joyce PR, Karam EG, Lee CK, Lellouch J, Lepine JP, Newman S, Rubio-Stipec M, Wells JE, Wickramaratne P, Wittchen HU, Yeh EK (1996) Cross-national epidemiology of major depression and bipolar disorder. JAMA 276:293–299

Williams NA, Close J, Giouzeli M, Crow TJ (2006) Accelerated evolution of *Protocadherin11X/Y*: A candidate gene-pair for cerebral asymmetry and language. Am J Med Genet (Neuropsychiatric Genet) 141B:623–633

Wollenberg R (1889) Ueber psychische infection. Archiv Psychiatr 20:62–88

Yakovlev PI, Rakic P (1966) Patterns of decussation of bulbar pyramids and distribution of pyramidal tracts on two sides of the spinal cord. Am Neurol Assoc 91:366–367

Yoshida K, Sugano S (1999) Identification of a novel protocadherin gene (*PCDH11*) on the human XY homology region in Xq21.3. Genomics 62:540–543

Microcephalies and DNA Repair

Edward C. Gilmore[1,2] and *Christopher A. Walsh*[1,3]

Head circumferences that are significantly smaller than normal are generally associated with lower intelligence, showing the relationship between intelligence and brain size (Dolk 1991). Some of the factors that lead to normal brain size are beginning to be understood by the study of genes that lead to small brains when disrupted. Some of the genes that cause microcephaly are integral to DNA repair. The relationships between different mechanisms of microcephaly are just beginning to be understood. Interestingly, there are potential relationships between the various mechanisms.

Microcephaly

The many etiologies of microcephaly generally can be categorized as environmental or genetic. Genetic causes of microcephaly include both degenerative and static diseases. Degenerative diseases present from shortly after birth to much later in life, with progressive worsening of symptoms. In contrast, abnormalities of primary brain development are present at birth. The clinical symptoms usually do not worsen over time. The abnormalities in brain development that lead to microcephaly will be the focus of this chapter.

The cerebral cortex makes up the largest structure of the human brain and is always affected in microcephaly. Even though most brain volume is made up of neuropil (glia, dendrites, etc.), the number of neurons determines the amount of neuropil present. As a consequence, the number of neurons is an important determinant of brain size. Knowing the factors regulating cerebral cortical cell number is critical to understanding the many causes of microcephaly. Because murine cerebral cortical development is best understood, it will be presented as a model, though higher mammals likely use similar mechanisms (Caviness et al. 1995). Proper regulation of this process is important since most cerebral cortex neurons can be born only during the normal neurogenesis process (Spalding et al. 2005; Bhardwaj et al. 2006).

[1] Department of Neurology, Beth Israel Deaconess Medical Center,
e-mail: ecgilmore@partners.org
[2] Child Neurology, Massachusetts General Hospital
[3] Division of Genetics, Children's Hospital Boston, Howard Hughes Medical Institute, Boston, MA, USA

Gage et al.
Retrotransposition, Diversity and the Brain
© Springer-Verlag Berlin Heidelberg 2008

Model of Cerebral Cortical Development

The murine cerebral cortex develops in the second half of gestation from a single layer of pseudostratified epithelium to a six-layered structure from embryonic day 10 (E10) to around birth. When neurons differentiate, they migrate away from the ventricular zone, next to the ventricular surface, towards the pial surface and past any previously born neurons to the superficial regions of the cortical plate. This process results in a cortex that has the earliest born cortical neurons residing in the deepest portions of the cerebral cortex whereas the latest born reside superficially. There are also functional differences between deep and superficial neurons, with the deepest neurons projecting to thalamic regions and below whereas more superficial neurons project to other regions of cerebral cortex. Prior to neurogenesis, the founding pool of precursor cells in the ventricular zone appears to be regulated via programmed cell death, as shown by the targeted deletion of genes required for apoptosis resulting in enlargement of the cerebral cortex from excessive precursor cells, not from lack of apoptosis later in development (Cecconi et al. 1998; Hakem et al. 1998; Kuida et al. 1998; Yoshida et al. 1998). The cell cycle length of dividing cells is critical because slower cell cycles result in fewer cell divisions within a given period of time. The length of neurogenesis determines how many cell cycles can occur.

The differentiating-to-proliferating ratios of ventricular precursors are also important to cerebral cortical size. As cells divide, there are three potential fates for the two daughter cells: two precursors, one precursor and one neuron, or two neurons. As neuronal precursors divide, the fates of the two precursors or two differentiated neurons can be thought of as symmetric in that both cells have the same fate. In contrast, when daughter cells are one differentiated neuron and one precursor, they can be thought of as asymmetric because the fates of the two cells are different. These choices are tightly controlled because if too many cells left the proliferating pool to differentiate early in cortical development, there would be fewer cells in later portions of neurogenesis, resulting in microcephaly (see Fig. 1B). Since the timing of differentiation determines laminar position and phenotype (McConnell and Kaznowski 1991), too many neurons differentiating early would leave too many early-born, deep neurons and too few later-born, superficial neurons.

Each factor that determines the number of neurons, progenitor pool size, cell cycle length, and differentiating/proliferating ratio can be affected by diseases causing microcephaly. Not surprisingly, chromosome number, growth factors, and transcription factors can play an important role in each of these events (Haydar et al. 2000; Hodge et al. 2004; Roy et al. 2004; Yun et al. 2004). Differentiation/proliferation ratios are likely disrupted in many microcephaly vera patients.

Specific Microcephalies

Microcephaly vera (true microcephaly) is a group of autosomal recessive diseases of abnormal brain development that result in mental retardation but not other neurological abnormalities (Woods et al. 2005). There are six known genetic loci that result in microcephaly vera (MCPH1-6); four of these have been cloned, but there are undoubtedly others. *MCPH5/ASPM*, *MCPH3/CDK5RAP2*, and *MCPH6/CENPJ* encode proteins that

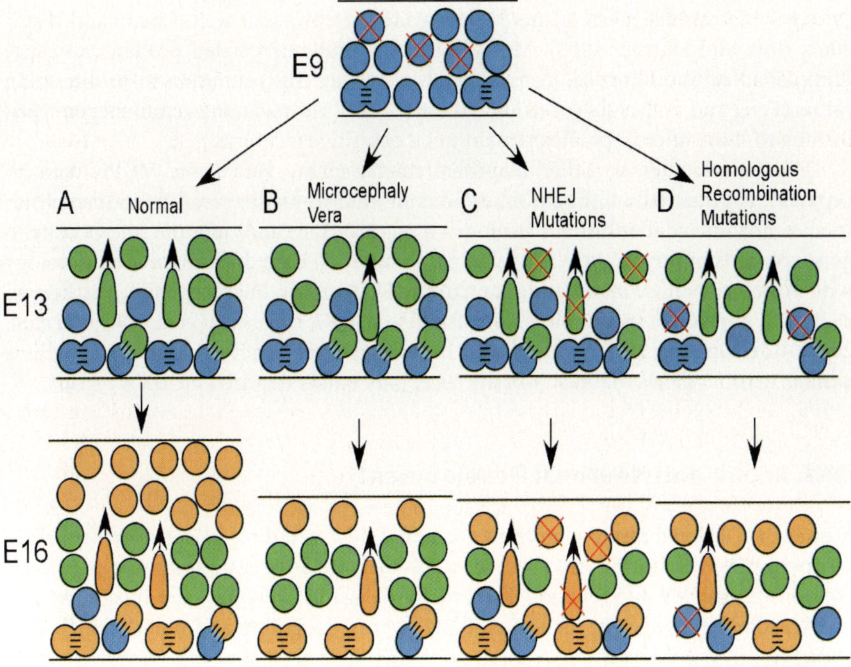

Fig. 1. Models of normal development and microcephaly. Prior to cerebral corticogenesis at E9, cell division produces more precursors via symmetric division (blue cells separating in the plane of the ventricular surface) whereas apoptosis (*crossed-out cells*) eliminates some, establishing the proper number of cerebral cortical precursors. (**A**) During normal development early in corticogenesis at E13, many symmetrically dividing ventricular zone cells produce two additional precursors (*blue cells*). However, asymmetric cell division (cleave plane away from ventricular surface) produces one neuron that will eventually reside in deep regions of cortex (*green*) and one precursor. Differentiating neurons migrate (*elongated cell with arrowhead*) to the superficial regions of the cortical plate. Late in corticogenesis at E16, fewer mitoses produce additional precursors, and more neurons (*tan circles*) are produced. Differentiating neurons migrate past earlier born neurons to superficial regions of the cortical plate. (**B**) During early cortical plate stage at E13, asymmetric division occurs too frequently, resulting in too many early-born, deep layer neurons and too few precursors within the ventricular zone. Too few neuroblasts later in development at E16 result in too few late-born, superficial neurons within the cortical plate since they are produced at this time, with microcephaly as a result. (**C**) Mutations in NHEJ genes result in cell death of differentiated neurons (*crossed-out cells*) that can be in either the cortical plate or deeper near the ventricular zone. This process results in microcephaly because too few neurons are present due to apoptosis after differentiation. (**D**) Mutations in XRCC2, which recent studies have shown are required for homologous recombination, appear to result in cell death of precursor cells and not of differentiated cells. Too few cortical neurons are present because too few precursors are present

likely associate with spindle poles/centrosomes (Bond et al. 2005; Kouprina et al. 2005). These genes potentially play a critical role in determining if neuronal precursors continue to divide or differentiate. In *Drosophila*, cellular polarity determines neuronal fate via asymmetric division of cellular components (see Wodarz 2005 for review). There is evidence that similar mechanisms play a role in vertebrates as well (Chenn and Walsh 2002; Gotz and Huttner 2005). Mutations in a spindle-associated protein can cause diminished cerebral cortical sizes through changing differentiation-to-proliferation ratios (Feng and Walsh 2004; see Fig. 1B for model). However, the actual mechanisms leading to these microcephalies remain to be definitively determined.

The function of the other identified microcephaly vera gene, MCPH1/micro-cephalin, is potentially different. Mutations in microcephalin result in abnormalities in chromosome condensation (Trimborn et al. 2004). In addition, the microcephalin gene contains multiple BRCA1 C-terminal domains. These domains share homology with proteins involved in cell cycle control and DNA repair. Microcephalin, also known as BRIT1, appears to play a role in the response to DNA repair (Xu et al. 2004; Lin et al. 2005; Alderton et al. 2006); which could indicate that the function of microcephalin is similar to other genes that result in microcephaly and is required for DNA repair.

DNA Repair and Neuronal Development

Damage to DNA can result in multiple abnormalities, including cancer, from activation of oncogenes, disruption of cell cycle inhibitors, or defects in apoptotic pathways. Neurons terminally differentiate, never to divide again. However, they are still subject to DNA-damaging agents, such as oxidative stress, radiation, and other insults. This susceptibility is especially important for humans since most, if not all, cortical neurons may be present at the time of birth (Au and Fishell 2006). If DNA is not repaired properly, the cells may undergo apoptosis or there could be defects in gene products.

Neurological phenotypes that result from abnormalities in DNA repair genes fall into two groups. Some cause neurodegenerative diseases and others cause abnormal-ities in neuronal development. These two phenotypes are not mutually exclusive, but there are potentially useful distinctions. Those that cause neurodegeneration are not the focus of this chapter and will only be discussed briefly to contrast with the devel-opmental mutations.

Most of the neurological degenerative diseases that involve DNA repair revolve around damage to the cerebellum and result in ataxia, a movement disorder. The best characterized of these diseases is ataxia telangiectasia (A-T), resulting from mutations in the ATM gene (Savitsky et al. 1995). Patients with A-T have progressive ataxia, abnormal eye movements, dermal blood vessel proliferation, cancer predisposition and immunodeficiency (Chun and Gatti 2004). A clinically similar disorder is known as A-T-like disease and has similar clinical features. It is caused by mutations in the gene MRE11 (Stewart et al. 1999). ATM is a protein that exists very early in the signaling cascade that responds to DNA damage, in double-stranded DNA in particular. MRE11 is a component of MRN complex that is thought to help activate ATM and it coordinates downstream signaling effects of ATM as well (Uziel et al. 2003; Taylor et al. 2004). Other ataxia syndromes, ataxia with oculomotor apraxia (AOA) 1 and spinocerebellar ataxia with neuropathy, are associated with mutations in the genes for aprataxin and tyrosyl-DNA phosphodiesterase 1 (TDP1; Moreira et al. 2001; Takashima et al. 2002). Aprataxin helps release adenosine mono-phosphate bound to 5' phosphates that can occur during ligation of broken DNA ends (Ahel et al. 2006; Rass et al. 2007). TDP1 is involved in single-stranded DNA repair, and its mechanism of action could be related to helping release topoisomerase that is covalently attached to DNA (El-Khamisy et al. 2005). Cockayne syndrome is associated with spasticity, mental retardation, and hearing loss that seems to affect the white matter (axon tracts) more than the gray matter (neurons; Rapin et al. 2000). It results from abnormalities in the single-stranded DNA repair pathway nucleotide excision repair (Fousteri et al. 2006). Xeroderma pigmentosum is characterized by photosensitivity, and cancer can also result in neuronal degeneration leading to microcephaly. Xeroderma pigmentosum has been associated with eight different genes involved in nucleotide-excision repair (Lehmann 2003).

In contrast to the degenerative diseases in which neuronal tissues form normally and eventually die, abnormalities in neuronal production characterize the developmental diseases. A number of DNA repair genes are required for normal neuronal development. Like microcephaly vera described above, too few neurons result in smaller brains. Nijmegen breakage syndrome is associated with microcephaly at birth, mental retardation, cancer, and immunodeficiency (Digweed and Sperling 2004). This syndrome results from mutations in the NBS1 gene (Carney et al. 1998; Matsuura et al. 1998; Varon et al. 1998). NBS1 is also part of the MRN complex that helps activate ATM and also acts downstream of ATM activation (see Kobayashi et al. 2004 for review). Seckel syndrome has primary dwarfism with even more profound microcephaly than is proportional to the body (Shanske et al. 1997). It has multiple linkage groups, but one form is caused by mutations in the ATM- and RAD3-related gene, ATR. ATR functions in a manner similar to ATM and helps activate cascades that respond to DNA damage (Niida and Nakanishi 2006). Seckle syndrome patients do not have cancer predisposition or immunodeficiency. Another disease characterized by microcephaly, mental retardation and immunodeficiency is associated with mutations in the DSB repair gene cernunnos (Buck et al. 2006). Cerunnos is involved in non-homologous end joining (NHEJ), which repairs double-stranded DNA breaks. NHEJ brings two damaged ends together, sometimes removing nucleotides before ligation and often resulting in imperfect repairs (Lees-Miller and Meek 2003). It can be used during all parts of the cell cycle (Rothkamm et al. 2003). Homologous recombination (HR) uses a replicated strand of DNA as a template for repair of double-strand breaks and does not introduce errors (Sonoda et al. 2006). HR can only be used in late S and G2 phases of cell replication since it requires a copied strand (Rothkamm et al. 2003).

Double-Stranded Break Repair and Microcephaly

NHEJ repair has a critical role during murine neuronal development as well. Targeted disruption of several genes required for NHEJ interrupts lymphocyte development and DNA repair and causes neuronal death whereas all other tissues are apparently spared. Disruption of Ligase IV, XRCC4, Ku70/Ku80, and some mutations in DNA-dependent kinase (DNA-PK) results in partial or severe neuronal cell death around the time of terminal differentiation in mice (Barnes et al. 1998; Frank et al. 1998; Gao et al. 1998; Gu et al. 2000; Vemuri et al. 2001). Disruption of ATM (a DNA damage sensor) or p53 (a downstream DNA damage mediator that helps regulate between cell cycle arrest and apoptosis) along with ligase IV or p53 with XRCC4 prevents neuronal death during development (Frank et al. 2000; Gao et al. 2000; Lee et al. 2000; Sekiguchi et al. 2001). Interestingly, deletion of ATM in addition to Ku70, Ku80 or DNA-PK results in early embryonic lethality for unknown reasons (Gurley and Kemp 2001; Sekiguchi et al. 2001). Human mutations in Ligase IV have also been found to result in microcephaly and immunodeficiency, supporting the relevance of murine findings in understanding human development (O'Driscoll et al. 2001).

HR is also critical for murine brain development. Targeted disruption of XRCC2, a gene involved in HR, results in apoptosis within the developing brain (Deans et al. 2000). Although these mice do not survive birth and fewer embryos than expected are present late in gestation, these authors found that the dying cells appeared to be differentiated neurons. However, a more recent analysis with conditional deletion of XRCC2 using a neuron-specific Cre found that dying cells appeared to be proliferative, not post-mitotic (Orii et al. 2006). The reason for the difference in findings is not clear at this point. Dividing precursors may rely upon HR to repair DNA damage while they are still proliferative since HR has the advantage that it is not error prone. However, once neurons differentiate, HR is no longer an option since HR requires a copied strand. Therefore, NHEJ must be used to repair double-stranded DNA breaks in differentiated neurons, which could explain why most dying cells in NHEJ mutants appear to be differentiated neurons.

For the purpose of this chapter, we have separated DNA repair genes into those that affect primary brain development from those that cause neurodegeneration. Interestingly, defects that either cause neuronal cell death during development (ligase VI, XRCC4, Ku70, Ku80, DNA-PK, NBS) or microcephaly (Seckles, cernunnos, ligase VI, Nijmegen breakage syndrome) are associated with double-stranded DNA break repair. Many mutations that result in neuronal degeneration are in genes that are critical for single-stranded DNA repair (ataxia with oculomotor apraxia 1, spinocerebellar ataxia with neuropathy, cockayne syndrome, and xeroderma pigmentosum). However, the genes mutated in A-T and A-T-like disorder play a role in double-stranded DNA repair. In fact, NBS1 (Nijmegen breakage syndrome) and MRE11 (A-T-like disorder) are within the same complex upstream of ATM and ATR but may have different functions in downstream signaling. The reason underlying the phenotypic differences of ATM/ATR and NBS1/MRE11 mutations remains a mystery. Part of this underlying mystery is the requirement for DNA repair in the first place (see below).

Integrating Models of Microcephaly

Different models for the development of microcephaly can be created from the data presented so far (Fig. 1). Microcephaly can develop from failure to produce enough neurons, which occurs when precursors cannot perform enough cell cycles or when ratios of differentiating-to-proliferating cells favor differentiation too early. *MCPH5/ASPM*, *MCPH3/CDK5RAP2*, and *MCPH6/CENPJ* fall into this category. Alternatively, the cellular death in differentiated neurons or precursors that results from mutations in DNA repair proteins will also result in microcephaly.

Data also suggest that the processes of DNA repair and neuronal symmetric/asymmetric division could be related. As discussed above, microcephalin potentially is involved in DNA repair and chromosome condensation, which could mean that microcephalin is involved in both models of microcephaly, DNA repair and cell proliferation. In addition, the identification of PAR3 as a protein that can work in conjunction with the NHEJ pathways suggests intriguing ties between DNA repair and cell polarity establishment (Fang et al. 2007). PAR3 is known to be involved in controlling asymmetric division in primitive organisms and establishing neuronal polarity in mammals (Joberty et al. 2000; Lin et al. 2000; Shi et al. 2003; Nishimura et al. 2004). However, PAR3 also is associated with the NHEJ proteins Ku70/80 in normal conditions and DNA-PK after irradiation. In addition, reduction of PAR3 causes cells to be more susceptible to irradiation (Fang et al. 2007). The fact that PAR3 has a role both in polarity and DNA repair could mean that DNA repair intrinsically is associated with neuronal differentiation related to asymmetric division. Alternatively, evolution may be merely resourceful in using the same factors for distinct pathways, both involving DNA, either in repairing DNA or translocating nuclei into a differentiated cell.

Sources of DNA Breaks During Development

The vast diversity found among neurons has long led to speculation that DNA rearrangements could play a role in neuronal development as it does in lymphoid cells. For instance, the restricted expression of one or several olfactory receptors in an individual olfactory neuron out of approximately one thousand present in the genome (Zhang and Firestein 2002; Niimura and Nei 2003) was a candidate for DNA rearrangement. Some hypothesized a regulatory mechanism for olfactory neurons that was analogous to immunoglobulin and T-cell receptor development. The expression of RAG1 in the nervous system further fueled speculation (Chun et al. 1991). However, targeted deletion in RAG1 did not result in any neuronal abnormalities (Mombaerts et al. 1992). In fact, olfactory neurons can be used for somatic nuclear transfer to clone mice with normal expression of olfactory receptors, indicating that there are no permanent changes to the genome in these cells (Eggan et al. 2004; Li et al. 2004). However, when cortical plate neurons were used as the source for nuclei in similar experiments, they performed poorly. Nuclei derived from the ventricular zone were 10 times more efficient at producing clones than differentiated neurons from the cortical plate (Yamazaki et al. 2001). The reason for the reduction in the cortical neurons totipotency remains unclear.

A potential mechanism for DNA damage during development is oxidation. Evidence of excessive oxidative damage can be found in neurons deficient in NHEJ repair proteins (Karanjawala et al. 2002; Narasimhaiah et al. 2005). The source of oxidative damage is speculated to be high energy metabolism during neuronal development associated with rapid proliferation. However, cell cycle lengths in embryonic cerebral cortex (9.5 hrs) are not dramatically faster than they are in the heart (13.4 hrs) at E10 in the rat. In addition, cell cycle times increase further into cerebral cortical development while cells become more susceptible to defects in NHEJ (Mirkes et al. 1989; Barnes et al. 1998; Frank et al. 1998; Gao et al. 1998; Gu et al. 2000; Vemuri et al. 2001; Nowakowski et al. 2002). However, differentiating cells must rely upon NHEJ to repair double-stranded breaks as discussed above, which may explain why neurons become more susceptible during development. In addition, neurons have a period of naturally occurring cell death whereas cardiac myocytes do not. This period of naturally occurring neuronal death matches the number of afferents to targets and prunes neurons that do not make proper connections (Oppenheim 1991). Therefore, neurons may be more susceptible to DNA damage because of priming for programmed cell death.

Another source of double-stranded breaks that could require repair during neuronal development is the movement of L1 retrotransposons (Muotri et al. 2005). The biology and functions of L1 retrotransposons are a main focus of this symposium and do not need to be extensively detailed. Briefly, L1 retrotransposons appear to be able to take advantage of DNA damage to find loci to integrate into (Rudin and Thompson 2001; Hagan et al. 2003). In addition, abnormalities in double-stranded break repair factors can facilitate movement of retrotransposons (Morrish et al. 2002). However, other retrotransposons may have difficulty moving in the absence of double-stranded DNA break repair (Downs and Jackson 1999; Izsvak et al. 2004). The movement of retrotransposons in developing neurons may require the machinery of DNA repair and their movement may be facilitated by other sources of DNA damage.

The requirements for proper neuronal development are beginning to be understood by study of both human and mouse mutations that result in microcephaly. The cellular processes that lead to deficient numbers can be caused by either failure to produce enough cells or cell death from failure to repair DNA. The relationship between these seemingly unrelated processes is interesting but remains circumstantial. Increasing neuronal diversity through movement of L1 retrotransposons or other mechanisms associated with DNA repair is intriguing but is not well defined. However, DNA repair is clearly integral to neuronal development, and understanding why that is so is likely to be beneficial for understanding human disease.

References

Ahel I, Rass U, El-Khamisy SF, Katyal S, Clements PM, PJ McKinnon PJ, Caldecott KW, West SC (2006) The neurodegenerative disease protein aprataxin resolves abortive DNA ligation intermediates. Nature 443:713–716

Alderton GK, Galbiati L, Griffith E, Surinya KH, Neitzel H, Jackson AP, Jeggo PA, O'Driscoll M (2006) Regulation of mitotic entry by microcephalin and its overlap with ATR signalling. Nature Cell Biol 8:725–733

Au E, Fishell EG (2006) Adult cortical neurogenesis: nuanced, negligible or nonexistent? Nature Neurosci 9:1086–1088

Barnes DE, Stamp G, Rosewell I, Denzel A, Lindahl T (1998) Targeted disruption of the gene encoding DNA ligase IV leads to lethality in embryonic mice. Curr Biol 8:1395–1398

Bhardwaj RD, Curtis MA, Spalding KL, Buchholz BA, Fink D, Bjork-Eriksson T, Nordborg C, Gage FH, Druid H, Eriksson PS, Frisen J (2006) Neocortical neurogenesis in humans is restricted to development. Proc Natl Acad Sci USA 103:12564–12568

Bond J, Roberts E, Springell K, Lizarraga SB, Scott S, Higgins J, Hampshire DJ, Morrison EE, Leal GF, Silva EO, SM Costa SM, Baralle D, Raponi M, Karbani G, Rashid Y, Jafri H, Bennett C, Corry P, Walsh CA, Woods CG (2005) A centrosomal mechanism involving CDK5RAP2 and CENPJ controls brain size. Nature Genet 37:353–355

Buck D, Malivert L, de Chasseval R, Barraud A, MC Fondaneche MC, Sanal O, Plebani A, Stephan JL, Hufnagel M, le Deist F, Fischer A, Durandy A, de Villartay JP, Revy P (2006) Cernunnos, a novel nonhomologous end-joining factor, is mutated in human immunodeficiency with microcephaly. Cell 124:287–299

Carney JP, Maser RS, Olivares H, Davis EM, Le Beau M, Yates JR 3rd, Hays L, Morgan WF, Petrini JH (1998) The hMre11/hRad50 protein complex and Nijmegen breakage syndrome: linkage of double-strand break repair to the cellular DNA damage response. Cell 93:477–486

Caviness VS Jr, Takahashi T, Nowakowski RS (1995) Numbers, time and neocortical neuronogenesis: a general developmental and evolutionary model. Trends Neurosci 18:379–383

Cecconi F, Alvarez-Bolado G, Meyer BI, Roth KA, Gruss P (1998) Apaf1 (CED-4 homolog) regulates programmed cell death in mammalian development. Cell 94:727–737

Chenn A, Walsh ACA (2002) Regulation of cerebral cortical size by control of cell cycle exit in neural precursors. Science 297:365–369

Chun HH, Gatti HRA (2004) Ataxia-telangiectasia, an evolving phenotype. DNA Repair (Amst) 3:1187–1196

Chun JJ, Schatz DG, Oettinger MA, Jaenisch R, Baltimore D (1991) The recombination activating gene-1 (RAG-1) transcript is present in the murine central nervous system. Cell 64:189–200

Deans B, Griffin CS, Maconochie M, Thacker J (2000) Xrcc2 is required for genetic stability, embryonic neurogenesis and viability in mice. Embo J 19:6675–6685

Digweed M, Sperling K (2004) Nijmegen breakage syndrome: clinical manifestation of defective response to DNA double-strand breaks. DNA Repair (Amst) 3:1207–1217

Dolk H (1991) The predictive value of microcephaly during the first year of life for mental retardation at seven years. Dev Med Child Neurol 33:974–983

Downs JA, Jackson SP (1999) Involvement of DNA end-binding protein Ku in Ty element retro-transposition. Mol Cell Biol 19:6260–6268

Eggan K, Baldwin K, Tackett M, Osborne J, Gogos J, Chess A, Axel R, Jaenisch R (2004) Mice cloned from olfactory sensory neurons. Nature 428:44–49

El-Khamis, SF, GM Saifi GM, Weinfeld M, Johansson F, Helleday T, Lupski JR, Caldecott KW (2005) Defective DNA single-strand break repair in spinocerebellar ataxia with axonal neuropathy-1. Nature 434:108–113

Fang L, Wang Y, Du D, Yang G, Tak Kwok T, Kai Kong S, Chen B, Chen DJ, Chen Z (2007) Cell polarity protein Par3 complexes with DNA-PK via Ku70 and regulates DNA double-strand break repair. Cell Res 17:100–116

Feng Y, Walsh CA (2004) Mitotic spindle regulation by Nde1 controls cerebral cortical size. Neuron 44:279–293

Fousteri M, Vermeulen W, van Zeeland AA, Mullenders LH (2006) Cockayne syndrome A and B proteins differentially regulate recruitment of chromatin remodeling and repair factors to stalled RNA polymerase II in vivo. Mol Cell 23:471–482

Frank KM, Sekiguchi JM, Seidl KJ, Swat W, Rathbun GA, Cheng HL, Davidson L, Kangaloo L, Alt FW (1998) Late embryonic lethality and impaired V(D)J recombination in mice lacking DNA ligase IV. Nature 396:173–177

Frank KM, Sharpless NE, Gao Y, Sekiguchi JM, Ferguson DO, Zhu C, Manis, Horner J, DePinho RA, Alt FW (2000) DNA ligase IV deficiency in mice leads to defective neurogenesis and embryonic lethality via the p53 pathway. Mol Cell 5:993–1002

Gao Y, Sun Y, Frank KM, Dikkes P, Fujiwara Y, Seidl KJ, Sekiguchi JM, Rathbun GA, Swat W, Wang J, Bronson RT, Malynn BA, Bryans M, Zhu C, Chaudhuri J, Davidson L, Ferrini R, Stamato T, Orkin SH, Greenberg ME, Alt FW (1998) A critical role for DNA end-joining proteins in both lymphogenesis and neurogenesis. Cell 95:891–902

Gao Y, Ferguson DO, Xie W, Manis JP, Sekiguchi J, Frank KM, Chaudhuri J, Horner J, DePinho RA, Alt FW (2000) Interplay of p53 and DNA-repair protein XRCC4 in tumorigenesis, genomic stability and development. Nature 404:897–900

Gotz M, Huttner WB (2005) The cell biology of neurogenesis. Nature Rev Mol Cell Biol 6:777–788

Gu Y, Sekiguchi J, Gao Y, Dikkes P, Frank K, Ferguson D, Hasty P, Chun J, Alt FW (2000) Defective embryonic neurogenesis in Ku-deficient but not DNA-dependent protein kinase catalytic subunit-deficient mice. Proc Natl Acad Sci USA 97:2668–2673

Gurley KE, Kemp CJ (2001) Synthetic lethality between mutation in Atm and DNA-PK(cs) during murine embryogenesis. Curr Biol 11:191–194

Hagan CR, Sheffield RF, Rudin CM (2003) Human Alu element retrotransposition induced by genotoxic stress. Nature Genet 35:219–220

Hakem R, Hakem A, Duncan GS, Henderson JT, Woo M, Soengas MS, Elia A, de la Pompa JL, Kagi D, Khoo W, Potter J, Yoshida R, Kaufman SA, Lowe SW, Penninger JM, Mak TW (1998) Differential requirement for caspase 9 in apoptotic pathways in vivo. Cell 94:339–352

Haydar TF, Nowakowski RS, Yarowsky PJ, Krueger BK (2000) Role of founder cell deficit and delayed neuronogenesis in microencephaly of the trisomy 16 mouse. J Neurosci 20:4156–4164

Hodge RD, D'Ercole AJ, O'Kusky JR (2004) Insulin-like growth factor-I accelerates the cell cycle by decreasing G1 phase length and increases cell cycle reentry in the embryonic cerebral cortex. J Neurosci 24:10201–10210

Izsvak Z, Stuwe EE, Fiedler D, Katzer A, Jeggo PA, Ivics Z (2004) Healing the wounds inflicted by sleeping beauty transposition by double-strand break repair in mammalian somatic cells. Mol Cell 13:279–290

Joberty G, Petersen C, Gao L, Macara IG (2000) The cell-polarity protein Par6 links Par3 and atypical protein kinase C to Cdc42. Nature Cell Biol 2:531–539

Karanjawala ZE, Murphy N, Hinton DR, Hsieh CL, Lieber MR (2002) Oxygen metabolism causes chromosome breaks and is associated with the neuronal apoptosis observed in DNA double-strand break repair mutants. Curr Biol 12:397–402

Kobayashi J, Antoccia A, Tauchi H, Matsuura S, Komatsu K (2004) NBS1 and its functional role in the DNA damage response. DNA Repair (Amst) 3:855–861

Kouprina N, Pavlicek A, Collins NK, Nakano M, Noskov VN, Ohzeki J, Mochida GH, Risinger JI, Goldsmith P, Gunsior M, Solomon G, Gersch W, Kim JH, Barrett JC, Walsh CA, Jurka J, Masumoto H, Larionov V (2005) The microcephaly ASPM gene is expressed in proliferating tissues and encodes for a mitotic spindle protein. Human Mol Genet 14:2155–2165

Kuida K, Haydar TF, Kuan CY, Gu Y, Taya C, Karasuyama H, Su MS, Rakic P, Flavell RA (1998) Reduced apoptosis and cytochrome c-mediated caspase activation in mice lacking caspase 9. Cell 94:325–337

Lee Y, Barnes DE, LindahL T, McKinnon PJ (2000) Defective neurogenesis resulting from DNA ligase IV deficiency requires Atm. Genes Dev 14:2576–2580

Lees-Miller SP, Meek K (2003) Repair of DNA double strand breaks by non-homologous end joining. Biochimie 85:1161–1173

Lehmann AR (2003) DNA repair-deficient diseases, xeroderma pigmentosum, Cockayne syndrome and trichothiodystrophy. Biochimie 85:1101–1111

Li J, Ishii T, Feinstein P, Mombaerts P (2004) Odorant receptor gene choice is reset by nuclear transfer from mouse olfactory sensory neurons. Nature 428:393–399

Lin D, Edwards AS, Fawcett JP, Mbamalu G, Scott JD, Pawson T (2000) A mammalian PAR-3-PAR-6 complex implicated in Cdc42/Rac1 and aPKC signalling and cell polarity. Nature Cell Biol 2:540–547

Lin SY, Rai R, Li K, Xu ZX, Elledge SJ (2005) BRIT1/MCPH1 is a DNA damage responsive protein that regulates the Brca1-Chk1 pathway, implicating checkpoint dysfunction in microcephaly. Proc Natl Acad Sci USA 102:15105–15109

Matsuura S, Tauchi H, Nakamura A, Kondo N, Sakamoto S, Endo S, Smeets D, Solder B, BH Belohradsky BH, Der Kaloustian VM, Oshimura M, Isomura M, Nakamura Y, Komatsu K (1998) Positional cloning of the gene for Nijmegen breakage syndrome. Nature Genet 19:179–181

McConnell SK, Kaznowski CE (1991) Cell cycle dependence of laminar determination in developing neocortex. Science 254:282–285

Mirkes PE, JL Ricks JL, Pascoe-Mason JM (1989) Cell cycle analysis in the cardiac and neuroepithelial tissues of day 10 rat embryos and the effects of phosphoramide mustard, the major teratogenic metabolite of cyclophosphamide. Teratology 39:115–120

Mombaerts P, Iacomini J, RS Johnson RS, Herrup K, Tonegawa S, Papaioannou VE (1992) RAG-1-deficient mice have no mature B and T lymphocytes. Cell 68:869–877

Moreira MC, Barbot C, Tachi N, Kozuka N, Uchida E, Gibson T, Mendonca P, Costa M, Barros J, Yanagisawa T, Watanabe M, Ikeda Y, Aoki M, Nagata T, Coutinho P, Sequeiros J, Koenig M (2001) The gene mutated in ataxia-ocular apraxia 1 encodes the new HIT/Zn-finger protein aprataxin. Nature Genet 29:189–193

Morrish TA, Gilbert N, Myers JS, Vincent BJ, Stamato TD, Taccioli GE, Batzer MA, Moran JV (2002) DNA repair mediated by endonuclease-independent LINE-1 retrotransposition. Nature Genet 31:159–165

Muotri AR, Chu VT, Marchetto MC, Deng W, Moran JV, Gage FH (2005) Somatic mosaicism in neuronal precursor cells mediated by L1 retrotransposition. Nature 435:903–910

Narasimhaiah R, Tuchman A, Lin SL, Naegele JR (2005) Oxidative damage and defective DNA repair is linked to apoptosis of migrating neurons and progenitors during cerebral cortex development in Ku70-deficient mice. Cereb Cortex 15:696–707

Niida, HM Nakanishi (2006) DNA damage checkpoints in mammals. Mutagenesis 21:3–9

Niimura Y, Nei M (2003) Evolution of olfactory receptor genes in the human genome. Proc Natl Acad Sci USA 100:12235–12240

Nishimura T, Kato K, Yamaguchi T, Fukata Y, Ohno S, Kaibuchi K (2004) Role of the PAR-3-KIF3 complex in the establishment of neuronal polarity. Nature Cell Biol 6:328–334

Nowakowski RS, Caviness VS Jr, Takahashi T, Hayes NL (2002) Population dynamics during cell proliferation and neuronogenesis in the developing murine neocortex. Results Probl Cell Differ 39:1–25

O'Driscoll M, Cerosaletti KM, Girard PM, Dai Y, Stumm M, Kysela B, Hirsch B, Gennery A, Palmer SE, Seidel J, Gatti RA, Varon R, Oettinger MA, Neitzel H, Jeggo PA, Concannon P (2001) DNA ligase IV mutations identified in patients exhibiting developmental delay and immunodeficiency. Mol Cell 8:1175–1185

Oppenheim RW (1991) Cell death during development of the nervous system. Annu Rev Neurosci 14:453–501

Orii KE, Lee Y, Kondo N, McKinnon PJ (2006) Selective utilization of nonhomologous end-joining and homologous recombination DNA repair pathways during nervous system development. Proc Natl Acad Sci USA 103:10017–10022

Rapin I, Lindenbaum Y, Dickson DW, Kraemer KH, Robbins JH (2000) Cockayne syndrome and xeroderma pigmentosum. Neurology 55:1442–1449

Rass U, Ahel I, WestSC (2007) Actions of aprataxin in multiple DNA repair pathways. J Biol Chem 282:9469–9474

Rothkamm K, Kruger I, Thompson LH, Lobrich M (2003) Pathways of DNA double-strand break repair during the mammalian cell cycle. Mol Cell Biol 23:5706–5715

Roy K, Kuznicki K, Wu Q, Sun Z, Bock D, Schutz G, Vranich N, Monaghan AP (2004) The Tlx gene regulates the timing of neurogenesis in the cortex. J Neurosci 24:8333–8345

Rudin CM, Thompson CB (2001) Transcriptional activation of short interspersed elements by DNA-damaging agents. Genes Chromosomes Cancer 30:64–71

Savitsky K, Bar-Shira A, Gilad S, Rotman G, Ziv Y, Vanagaite L, Tagle DA, Smith S, Uziel T, Sfez S, Ashkenazi M, Pecker I, Frydman M, Harnik R, Patanjali SR, Simmons A, Clines GA, Sartiel A, Gatti RA, Chessa L, Sanal O, Lavin MF, Jaspers NG, Taylor AM, Arlett CF, Miki T, Weissman SM, Lovett M, Collins FS, Shiloh Y (1995) A single ataxia telangiectasia gene with a product similar to PI-3 kinase. Science 268:1749–1753

Sekiguchi J, DO Ferguson DO, Chen HT, Yang EM, Earle J, Frank K, Whitlow S, Gu Y, Xu Y, Nussenzweig A, Alt FW (2001) Genetic interactions between ATM and the nonhomologous end-joining factors in genomic stability and development. Proc Natl Acad Sci USA 98:3243–3248

Shanske A, Caride DG, Menasse-Palmer L, Bogdanow A, Marion RW (1997) Central nervous system anomalies in Seckel syndrome: report of a new family and review of the literature. Am J Med Genet 70:155–158

Shi SH, Jan LY, Jan YN (2003) Hippocampal neuronal polarity specified by spatially localized mPar3/mPar6 and PI 3-kinase activity. Cell 112:63–75

Sonoda E, Hochegger H, Saberi A, Taniguchi Y, Takeda S (2006) Differential usage of non-homologous end-joining and homologous recombination in double strand break repair. DNA Repair (Amst) 5:1021–1029

Spalding KL, Bhardwaj RD, Buchholz BA, Druid H, Frisen J (2005) Retrospective birth dating of cells in humans. Cell 122:133–143

Stewart GS, Maser RS, Stankovic T, Bressan DA, Kaplan MI, Jaspers NG, Raams A, Byrd PJ, Petrini JH, Taylor AM (1999) The DNA double-strand break repair gene hMRE11 is mutated in individuals with an ataxia-telangiectasia-like disorder. Cell 99:577–587

Takashim, H, Boerkoel CF, John J, Saifi GM, Salih MA, Armstrong D, Y Mao, Quiocho FA, Roa BB, Nakagawa M, Stockton DW, Lupski JR (2002) Mutation of TDP1, encoding a topoisomerase I-dependent DNA damage repair enzyme, in spinocerebellar ataxia with axonal neuropathy. Nature Genet 32:267–272

Taylor AM, Groom A, Byrd PJ (2004) Ataxia-telangiectasia-like disorder (ATLD) - its clinical presentation and molecular basis. DNA Repair (Amst) 3:1219–1225

Trimborn M, Bell SM, Felix C, Rashid Y, Jafri H, Griffiths PD, Neumann LM, Krebs A, Reis A, Sperling K, Neitzel H, Jackson AP (2004) Mutations in microcephalin cause aberrant regulation of chromosome condensation. Am J Human Genet 75:261–266

Uziel T, Lerenthal Y, Moyal L, Andegeko Y, Mittelman L, Shiloh Y (2003) Requirement of the MRN complex for ATM activation by DNA damage. Embo J 22:5612–5621

Varon R, Vissinga C, Platzer M, Cerosaletti KM, Chrzanowska KH, Saar K, Beckmann G, Seemanova E, Cooper PR, Nowak NJ, Stumm M, Weemaes CM, Gatti RA, Wilson RK, Digweed M, Rosenthal A, Sperling K, Concannon P, Reis A (1998) Nibrin, a novel DNA double-strand break repair protein, is mutated in Nijmegen breakage syndrome. Cell 93:467–476

Vemuri MC, Schiller E, Naegele JR (2001) Elevated DNA double strand breaks and apoptosis in the CNS of scid mutant mice. Cell Death Differ 8:245–255

Wodarz A (2005) Molecular control of cell polarity and asymmetric cell division in Drosophila neuroblasts. Curr Opin Cell Biol 17:475–481

Woods CG, Bond J, Enard W (2005) Autosomal recessive primary microcephaly (MCPH): a review of clinical, molecular, and evolutionary findings. Am J Human Genet 76:717–728

Xu X, Lee J, Stern DF (2004) Microcephalin is a DNA damage response protein involved in regulation of CHK1 and BRCA1. J Biol Chem 279:34091–34094

Yamazaki Y, Makino H, Hamaguchi-Hamada K, Hamada S, Sugino H, Kawase E, Miyata T, Ogawa M, Yanagimachi R, Yagi T (2001) Assessment of the developmental totipotency of neural cells in the cerebral cortex of mouse embryo by nuclear transfer. Proc Natl Acad Sci USA 98:14022–14026

Yoshida H, Kong YY, Yoshida R, Elia AJ, Hakem A, Hakem R, Penninger JM, Mak TW (1998) Apaf1 is required for mitochondrial pathways of apoptosis and brain development. Cell 94:739–750

Yun K, Mantani A, Garel S, Rubenstein J, Israel MA (2004) Id4 regulates neural progenitor proliferation and differentiation in vivo. Development 131:5441–5448

Zhang X, Firestein S (2002) The olfactory receptor gene superfamily of the mouse. Nature Neurosci 5:124–133

Subject Index

Antibodies (diversity of)   54
Antibodies to retroviral proteins in psychiatric
   disorders   72–76
Asymmetry as a determinant of human ability
   94, 95

Brain complexity   53–64

Catalytic RNA   2–4, 34, 35
Cerebral asymmetries   92–96
Cerebral cortical development   109–112
Cerebral dominance gene   87–108
Crick F.   53
Cytomegalovirus   67

Dawkins R.   25
Diversity (Generation of)   53–64
DNA breaks during development   115, 116
DNA repair and neuronal development   112,
   113
Double-stranded break repair and
   microcephaly   113, 114

Entropy exchange model   36

Gag protein   36
Gene creation   44, 45
Glial cell missing (Gcm)   71, 72

Herpes simplex virus   67
Hippocampal learning (molecular and circuit
   mechanisms for)   13–18
HIV   36, 67
Human diversity and L1 retrotransposon
   biology   43–52
Human endogenous retrovirus   67–80
Human evolution   87, 92–94

Influenza virus   67, 72, 88, 89

Junk DNA   53, 61

Language lateralization   94

LINE-1   2, 22–28, 34, 43–62, 69, 116
– Evolutionary consequences of L1 impact in
   neuronal genomes   60–62
– Mechanism of LINE-1 integration   24
– Replicative cycle of LINE-1   23

Manic depressive illness   87, 88, 92
McClintock B.   8, 53
Memory   13–18
Microcephaly and DNA repair   109–120
Mobile element   8, 55
Multiple sclerosis   69

Neural stem cell   8, 9, 56–58
Neuronal progenitor   53, 58–60
Neuropsychiatric disorders   65–108
NMDA receptor   13–17
Novelty detection   17

Pattern completion   13–15
Pattern separation   15–17
Phylogenetic tree of life and rDNA sequence
   alignments   21
Polymorphisms in HERV K18   76–80
Prebiotic world   34, 40
Primitive RNA world   33–40
Prion PrP   39, 93
Protocadherin X/Y   87, 98
Proto-oncogene   92, 93
Psychosis   65–108
– Continuum of   91, 92
– Season of birth effect   92

RNA chaperone   35–40, 43
RNA world and brain complexity   53–64
RNA world hypothesis   33
Retrotransposition   8, 9
– Neural-specific retrotransposition   28
Retrotransposons   21–62, 98–103
– Ancient retrotransposons as possible
   remnants of the primitive RNA world
   33–42

– Cancer and   28, 55
– Creation of new genes and   43–52
– Human diversity and   43–52
– Mapping transposon insertion site   22, 27
– Natural and synthetic   21–32
– Neural-specific   28
– Psychosis and   87–108
– Replication   37–39
– Silencing and activation of   55–58
– Xq21.3/Yp and   98–103
Retrovirus   34, 40, 65–108
Reverse transcriptase   1–3, 8, 9, 24, 43, 92

Schizophrenia   65–108
– As an anomaly of development   95, 96
– Environmental theory of   87, 88
– Epidemiology   89, 90
– Genetic predisposition of   66, 88, 91–94
– Infectious disease and   88, 89
– Seasonality of   89

– Temporal geographical variations   89, 90
Selfish gene   25, 26, 53, 55
Sox protein   56–58
Syncitin   71, 72

Telomerase   1–12
– Control of   4–5
– Evolution of   2
– Reverse transcription and   8, 9
– Roles of   5, 6
– Stress and   7, 8
Telomeres and aging   6–8
Telomere and stress   7, 8
Telomeres and telomerases in human health
    and disease   1–12
Toxoplasma   67, 68

Xq21.3/Yp translocation   96–103
XY homology   96–98

List of volumes published in the series "Research and Perspectives in Neurosciences"

P. Ascher, D.W. Choi, Y. Christen (Eds.) (1991) Glutamate, Cell Death and Memory, ISBN 3-540-54134-9

F.H. Gage, Y. Christen (Eds.) (1992) Gene Transfer and Therapy in the Nervous System, ISBN 3-540-55889-6

A.-M. Thierry, J. Glowinski, P.S. Goldman-Rakic, Y. Christen (Eds.) (1994) Motor and Cognitive Functions of the Prefrontal Cortex, ISBN 3-540-57128-0

G. Buzsáki, R. Llinás, W. Singer, A Berthoz, Y. Christen (Eds.) (1994) Temporal Coding in the Brain, ISBN 3-540-58074-3

A.R. Damasio, H. Damasio, Y. Christen (eds.) (1996) Neurobiology of Decision-Making, ISBN 978-3-540-60143-2

F.H. Gage, Y. Christen (Eds.) (1997) Isolation, Characterization and Utilization of Stem Cells, ISBN 978-3-540-26253-4

A.M. Galaburda, Y. Christen (Eds.) (1997) Normal and Abnormal Development of the Cortex

J. Grafman, Y. Christen (Ed.) (1999) Neuronal Plasticity: Building a Bridge from the Laboratory to the Clinic, ISBN 978-3-540-64357-9

P. Patterson, C. Kordon, Y. Christen (Eds.) (1999) Neuro-Immune Interactions in Neurology and Psychiatric Disorders, ISBN 978-3-540-66013-2

C.E. Henderson, D. Green, J. Mariani, Y. Christen (Eds.) (2001) Neuronal Death by Accident or by Design, ISBN 978-3-540-41777-4

J. Mallet, Y. Christen (Eds.) (2003) Neurosciences at the Postgenomic Era, ISBN 978-3-540-00194-2

F.H. Gage, A. Björklund, A. Prochiantz, Y. Christen (Eds.) (2004) Stem Cells in the Nervous System: Functional and Clinical Implications, ISBN 978-3-540-20558-6

J.-P. Changeux, A.R. Damasio, W. Singer, Y. Christen (Eds.) (2005) Neurobiology of Human Values, ISBN 978-3-540-60143-2

B. Bontempi, A.J. Silva, Y. Christen (Eds.) (2007) Memories: Molecules and Circuits, ISBN 978-3-540-45698-8

Printing: Krips bv, Meppel, The Netherlands
Binding: Stürtz, Würzburg, Germany